APPLYING MATHEMATICS:
A Course in Mathematical Modelling

ELLIS HORWOOD SERIES IN MATHEMATICS AND ITS APPLICATIONS

Series Editor: Professor G. M. BELL, Chelsea College, University of London
(and within the same series)

Statistics and Operational Research

Editor: B. W. CONOLLY, Chelsea College, University of London

Baldock, G. R. & Bridgeman, T.	**MATHEMATICAL THEORY OF WAVE MOTION**
Beaumont, G. P.	**INTRODUCTORY APPLIED PROBABILITY**
Burghes, D. N. & Wood, A. D.	**MATHEMATICAL MODELS IN SOCIAL MANAGEMENT AND LIFE SCIENCES**
Burghes, D. N.	**MODERN INTRODUCTION TO CLASSICAL MECHANICS AND CONTROL**
Burghes, D. N. & Graham, A.	**CONTROL AND OPTIMAL CONTROL**
Burghes, D. N., Huntley, I. & McDonald, J.	**APPLYING MATHEMATICS**
Butkovskiy, A. G.	**GREEN'S FUNCTIONS AND TRANSFER FUNCTIONS HANDBOOK**
Butkovskiy, A. G.	**STRUCTURE OF DISTRIBUTED SYSTEMS**
Chorlton, F.	**TEXTBOOK OF DYNAMICS**
Chorlton, F.	**VECTOR AND TENSOR METHODS**
Conolly, B.	**TECHNIQUES IN OPERATIONAL RESEARCH Vol. 1: QUEUEING SYSTEMS Vol. 2: MODELS, SEARCH, RANDOMIZATION**
Dunning-Davies, J.	**MATHEMATICAL METHODS FOR MATHEMATICS, PHYSICAL SCIENCE AND ENGINEERING**
Eason, G., Coles, C. W., Gettinby, G.	**MATHEMATICS FOR THE BIOSCIENCES**
Exton, H.	**HANDBOOK OF HYPERGEOMETRIC INTEGRALS**
Exton, H.	**MULTIPLE HYPERGEOMETRIC FUNCTIONS**
Faux, I. D. & Pratt, M. J.	**COMPUTATIONAL GEOMETRY FOR DESIGN AND MANUFACTURE**
Goult, R. J.	**APPLIED LINEAR ALGEBRA**
Graham, A.	**KRONECKER PRODUCTS AND MATRIX CALCULUS: WITH APPLICATIONS**
Graham, A.	**MATRIX THEORY AND APPLICATIONS FOR ENGINEERS AND MATHEMATICIANS**
Griffel, D. H.	**APPLIED FUNCTIONAL ANALYSIS**
Hoskins, R. F.	**GENERALISED FUNCTIONS**
Hunter, S. C.	**MECHANICS OF CONTINUOUS MEDIA, 2nd (Revised) Edition**
Huntley, I. & Johnson, R. M.	**LINEAR AND NON-LINEAR DIFFERENTIAL EQUATIONS**
Jones, A. J.	**GAME THEORY**
Kemp, K. W.	**COMPUTATIONAL STATISTICS**
Kosinski, W.	**FIELD SINGULARITIES AND WAVE ANALYSIS IN CONTINUUM MECHANICS**
Marichev, O. I.	**INTEGRALS OF HIGHER TRANSCENDENTAL FUNCTIONS**
Meek, B. L. & Fairthorne, S.	**USING COMPUTERS**
Muller-Pfeiffer, E.	**SPECTRAL THEORY OF ORDINARY DIFFERENTIAL OPERATORS**
Nonweiler, T. R. F.	**COMPUTATIONAL MATHEMATICS: An Introduction to Numerical Analysis**
Oliviera-Pinto, F.	**SIMULATION CONCEPTS IN MATHEMATICAL MODELLING**
Oliviera-Pinto, F. & Conolly, B. W.	**APPLICABLE MATHEMATICS OF NON-PHYSICAL PHENOMENA**
Scorer, R. S.	**ENVIRONMENTAL AERODYNAMICS**
Smith, D. K.	**NETWORK OPTIMISATION PRACTICE: A Computational Guide**
Stoodley, K. D. C., Lewis, T. & Stainton, C. L. S.	**APPLIED STATISTICAL TECHNIQUES**
Sweet, M. V.	**ALGEBRA, GEOMETRY AND TRIGONOMETRY FOR SCIENCE STUDENTS**
Temperley, H. N. V. & Trevena, D. H.	**LIQUIDS AND THEIR PROPERTIES**
Temperley, H. N. V.	**GRAPH THEORY AND APPLICATIONS**
Twizell, E. H.	**COMPUTATIONAL METHODS FOR PARTIAL DIFFERENTIAL EQUATIONS IN BIOMEDICINE**
Whitehead, J. R.	**THE DESIGN AND ANALYSIS OF SEQUENTIAL CLINICAL TRIALS**

APPLYING MATHEMATICS: A Course in Mathematical Modelling

DAVID BURGHES, B.Sc., Ph.D., F.R.A.S., F.I.M.A.
Professor, School of Education,
University of Exeter

IAN HUNTLEY, B.A., Ph.D., F.I.M.A.
and
JOHN McDONALD, B.Sc., M.Sc., A.F.I.M.A.
Senior Lecturers
Paisley College of Technology, Scotland

ELLIS HORWOOD LIMITED
Publishers · Chichester

Halsted Press: a division of
JOHN WILEY & SONS
New York · Brisbane · Chichester · Toronto

First published in 1982 by

ELLIS HORWOOD LIMITED
Market Cross House, Cooper Street, Chichester, West Sussex, PO19 1EB, England

The publisher's colophon is reproduced from James Gillison's drawing of the ancient Market Cross, Chichester.

Distributors:

Australia, New Zealand, South-east Asia:
Jacaranda-Wiley Ltd., Jacaranda Press,
JOHN WILEY & SONS INC.,
G.P.O. Box 859, Brisbane, Queensland 40001, Australia

Canada:
JOHN WILEY & SONS CANADA LIMITED
22 Worcester Road, Rexdale, Ontario, Canada.

Europe, Africa:
JOHN WILEY & SONS LIMITED
Baffins Lane, Chichester, West Sussex, England.

North and South America and the rest of the world:
Halsted Press: a division of
JOHN WILEY & SONS
605 Third Avenue, New York, N.Y. 10016, U.S.A.

British Library Cataloguing in Publication Data
Burghes, D. N.
Applying mathematical modelling. –
(Ellis Horwood series in mathematics and its applications)
1. Mathematical models
I. Title II. Huntley, Ian
III. McDonald, John
511'.8 QA401

Library of Congress Card No. 82-6251 AACR2

ISBN 0-85312-417-5 (Ellis Horwood Ltd. – Library Edn.)
ISBN 0-85312-456-6 (Ellis Horwood Ltd. – Student Edn.)
ISBN 0-470-27523-5 (Halsted Press)

Typeset in Press Roman by Ellis Horwood Ltd.
Printed in Great Britain by Butler & Tanner, Frome, Somerset.

Table of Contents

Introduction

1. MATHEMATICAL MODELLING

Over the past few decades, there has been a considerable increase in the use of mathematical analysis, both for solving everyday problems and for theoretical developments of many disciplines. For example, economics, biology, geography and medicine have all seen a considerable increase in the use of quantitative techniques. Twenty years ago *applied mathematics* meant the application of mathematics to problems in mechanics and little else – now, applied mathematics, or as many people prefer to call it, applicable mathematics, could refer to the use of mathematics in many varied areas. The one unifying theme that these applications have is that of *mathematical modelling,* by which we mean the construction of a mathematical model to describe the situation under study. This process of changing a real-life problem into a mathematical one is not at all easy, we hasten to add, although one of the overall aims of this book is to improve your ability as a mathematical modeller.

We will soon see that it is not at all easy in fact even to make precise definitions about mathematical modelling, and many experts in the area are not fully in agreement. What, though, we want to emphasise here is not definitions but the practical nature of mathematics. We hope that we will not only improve your ability in using mathematics to solve practical problems, but also convince you that mathematics is a vitally relevant discipline, very much needed in the search for solutions to today's urgent problems.

2. INTRODUCTION TO THIS BOOK

There have been many books written during the past decade on the topic of mathematical modelling; all these books have been devoted to explaining and developing mathematical models, but very little space has been given to *how* to construct mathematical models, that is, how to take a real problem and convert it into a mathematical one. Although we appreciate that we might not yet have the best methods for teaching how to tackle real problems, we do at least regard this mastery of model formulation as a crucial step, and much of this book is devoted to attempting to make you more proficient in this process.

Our basic concept is that applied mathematicians become better modellers through more and more experience of tackling real problems. So in order to get the most out of this book, we stress that you must make a positive effort to tackle the many problems posed before looking at the solutions we have given. To help you to gain confidence in the art of modelling we have divided the book into four distinct sections.

Part I

In this section we describe three different examples of how mathematical analysis has been used to solve practical problems. These are all true accounts of how mathematical analysis has helped to provide solutions. We are not expecting you to do much at this stage, except to read through the case studies carefully, paying particular attention to the way in which the problems have been tackled – the process of translating the problem into a mathematical one.

We hope also that these case studies will convince you that mathematics can be of great use in solving important practical problems.

Part II

This section consists of a series of real problems, together with possible solutions and related problems. Each problem has a clear statement, and we very much encourage you to try to solve these problems in the first place *without* looking at the solutions we have given. The problems require for solution different levels of mathematics, and you might find that you have not yet covered some of the mathematical topics required. In general we have tried to order them, so that the level of mathematics required in the solutions increases as you move through the problems. Remember that we are only giving *our* solutions and, particularly if you don't look at our solution, you might well have a completely different approach which might provide a better solution.

If at this stage you are finding it very tough going, then we suggest you work through the solutions to the problems as in Part I, but at least have an attempt at the related problems set. These usually do *not* include any vastly new modelling strategy, but just a slight extension of the model already described.

We are convinced that the best way to become confident at using mathematics in practical situations is to gain more and more experience and consequently more mathematical maturity. This is why we have used this section to give you plenty of practice at attempting to solve real problems, when solutions are readily available if you do need help.

As in Part I, you should also analyse the approaches taken to solve practical problems. Although it will be apparent that different approaches work in different situations, it might also be apparent that there are some general methods and concepts in the model formation.

Part III

Here, at last, we try to give you some advice as to how to approach the tackling of real problem solving, and we give some general concepts involved in mathematical modelling. It must, though, again be stressed that we are all convinced that *experience* is the all-important ingredient needed for confidence in model formulation. If you have just read Parts I and II without making at least attempts at your *own* solutions to some of the problems set, you will not have gained any real experience in tackling real problems, and this section will not really be of much help. On the other hand, if you have taken the problem solving seriously in Part II, you might find the general advice given here helpful.

Part IV

Provided you have gained some sonfidence in tackling real problem solving in the earlier parts, you will be able to dabble with those problems in this section which appeal to you. Don't feel you must work systematically through this section, but look for problems you *want* to solve – these are the ones that you will have most success in solving.

Although we have not provided any solutions for this section, we do, though, give references where solutions can be found. We would again stress, though, that if you jump straight to the solutions, you will have gained little expertise in modelling. We hope that you might now be certain that mathematical analysis is of great importance, but we hope that you will aim further and attempt to become proficient in using mathematics to solve practical problems.

One of the great shortcomings of our educational system is that we are very good at training students to become proficient in mastering mathematical techniques, but we do little to show them how to use these techniques in practical situations. If we are really to find solutions to the many varied problems of society today, we must become experts at solving practical problems.

We hope that this book will at least point you in this direction. We are aware that this is not a finalised precise sort of text, but then using mathematics in practical problem solving is not a precise art. It is full of pitfalls and difficulties; but do not despair, you will find great excitement and satisfaction when you have had your first success at real problem solving!

Part I

Mathematics and Society

INTRODUCTION TO PART I

In this section of the book we give three examples of how mathematics has been of use in solving problems for which society has clearly required answers. Our main aim in this section is to give evidence that mathematics is a useful discipline, although we would also like you to look at the way in which mathematics has been introduced to solve these problems. By and large, though, we hope you will just read through these examples and enjoy them and gain motivation for a study of the rest of the book.

1 KING ARTHUR'S ROUND TABLE

1.1 Introduction to the Round Table

The picture opposite shows the round table in the great hall in Winchester Castle. The table top is 18 feet in diameter and divided into 25 sections, one for the King and places for 24 knights. Since the shape was *round*, no knight could assume precedence over others.

The first mention of a Round Table is in the *Roman de Brut* by the Norman poet Wace [1]. According to Wace, the Bretons told many stories about this table. There is a story about violent quarrels amongst the knights, because there were numerous tables and all wanted to be served at the most honourable. According to the legend, a Cornish carpenter then made Arthur a table at which 1600 men could sit, but which he could nevertheless carry with him!

There has been speculation about the origin of the Winchester round table. Was it really Arthur's round table? The existence of a round table at Winchester in the 15th century is confirmed by William Caxton in 1485, whilst John Harding says in his *Chronicle* (1484) that the Round Table "began at Winchester and ended at Winchester, and there it hangs still".

To put an end finally to the specualtion regarding the Winchester round table, in 1976 it was taken down from the wall on which it was fixed and a series of tests were employed in order to determine the date of origin.

1.2 Carbon Dating

One of the most crucial methods used to date the table was 'carbon dating', which had been developed in the late 1940s by an American chemist, W. F. Libby. (For this work he received the Nobel Prize in Chemistry in 1960.)

Carbon dating is based on the phenomenon of radioactivity, discovered and developed at the beginning of this century by Rutherford and others. Certain atoms are inherently *unstable*, so that after some time and without any outside influence they will undergo transition to an atom of a different element whilst emitting radiation. From experimental evidence, Rutherford formulated a simple and elegant model to describe the way in which a radioactive sample decays. If $N = N(t)$ represents the number of atoms in a radioactive sample at time t, then $\mathrm{d}N/\mathrm{d}t$ represents the rate of change of the number of atoms in the sample. Hence in *unit* time (assuming constant rate of change) the number of disintegrations will be $-\mathrm{d}N/\mathrm{d}t$.

Rutherford showed that the number of disintegrations was actually proportional to the number of atoms present; if the sample size were doubled, the number of disintegrations would double. Hence, we can write

$$\mathrm{d}N/\mathrm{d}t = -\lambda N \tag{1.1}$$

where λ is a positive constant, known as the **decay constant**. This constant has different values for different substances – the larger the value, the quicker the sample decays. Values for λ have to be found experimentally, although in practice we work with the **half-life**, τ, rather than λ. This is defined as the time taken for half a given quantity of atoms to decay, so that if $N = N_0$ say at $t = 0$, then $N = N_0/2$ at $t = \tau$, the half-life. To find the relationship between λ and τ, we first note that (1.1) has solution

$$N = N_0 e^{-\lambda t} \ . \tag{1.2}$$

Thus at time τ we have

$$\tfrac{1}{2}N_0 = N_0 e^{-\lambda\tau} \ ,$$

which gives

$$\frac{\ln 2}{\tau} \ .$$

The table below gives some examples of half-lives of various substances.

Substance	τ, half-life
Xenon 133	5 days
Barium 140	13 days
Lead 210	22 years
Strontium 90	25 years
Carbon 14	5568 years
Plutonium	23 103 years
Uranium 238	4.5×10^9 years

So for Carbon 14, an isotope of Carbon, the decay constant has value

$$\lambda = \frac{\ln 2}{\tau} = 1.245 \times 10^{-4} \text{ per year,}$$

using the half-life above.

We can now move onto the application in carbon dating. The earth's atmosphere is continuously bombarded by cosmic rays. This produces neutrons in the atmosphere, and these combine with nitrogen to form Carbon 14 (C^{14}). In living plants and animals, the rate of absorption of C^{14} is balanced by the natural decay, and an equilibrium state is reached. The basic assumption in carbon dating is that the intensity of bombardment of the earth's surface by cosmic rays has remained constant throughout time. This means that the original rate of disintegrations of C^{14} in a sample of wood is the same as the rate measured today in living wood. When the sample is formed (e.g. the table made) the wood is isolated from its original environment and the C^{14} atoms decay without any further absorption.

Suppose the table was formed at time $t = 0$. Let R_0 denote the original rate of disintegrations, so that

$$R_0 = [-\mathrm{d}N/\mathrm{d}t]_{t=0} = \lambda N_0 \ .$$

The present rate of disintegrations, $R(t)$, is given by

$$R(t) = -\,\mathrm{d}N/\mathrm{d}t = \lambda N = \lambda e^{-\lambda t} N_0$$

using (1.2). Hence $R_0/R(t) = e^{\lambda t}$, and, taking logs, we have

$$t = (1/\lambda) \ln|\{R_0/R(t)\} \qquad (1.3)$$

Thus, provided we can measure today's rate, $R(t)$, and we know the original R_0 (we assume that this has remained constant and so has the same value in living wood today), we can use (1.3) to compute the sample's age.

1.3 Application to Winchester's Round Table

For living wood, the rate of disintegrations (per minute per gram of sample) is 6.68, so we take this as the value when the table was formed, i.e., $R_0 = 6.68$. From measurements taken in 1977 from the table, the present rate is given by $R(t) = 6.08$. From (1.3), the age of the table is given by

$$t = \frac{1}{1.245 \times 10^{-4}} \quad \ln \frac{6.68}{6.08} = 756 \text{ years.}$$

This gives a date for the table of about AD 1220, which clearly indicates that the table was not King Arthur's (he lived in the 5th century). Further details of carbon dating are given in references [2] and [3].

REFERENCES

[1] Wace, (1155), *Roman de Brut,* ed. I. Arnold, (1938-1940), Paris.

[2] Maki, D. P. & Thompson, M., (1973), *Mathematical models and applications,* Prentice-Hall.

[3] Braun, M., (1975), *Differential equations and their applications,* Springer Verlag.

2 AIRCRAFT COLLISIONS

2.1 Summary

In the 1960s a controversy arose in the United States over plans to decrease the separation distance between aircraft crossing the North Atlantic. The aircraft owners, for reasons of economy, wanted to decrease the minimum separation distance, while the pilots felt this was unsafe and refused to operate the new system. In order to break this deadlock the International Civil Aviation Organisation (ICAO) set up a System Planning Group to collect and examine evidence and advise accordingly. This chapter describes, in very general terms, how the controversy was resolved and *both* parties satisfied.

[This chapter is based on an article by R. E. Machol entitled 'An aircraft collision model' which appeared in *Management Science,* **21**(10), 1975.]

2.2 Introduction

During the mid 1960s there were approximately 100 000 aircraft crossings of the North Atlantic each year. Most of the eastbound traffic left the New York area in the evening and arrived in northern Europe the next morning, while the westbound traffic left the London area in the afternoon and arrived in the northern part of the United States the same evening. The bulk of the traffic followed the same great-circle route, since this route was shortest and therefore economised on aircraft fuel and staff duty time. Consequently a considerable number of aircraft were simultaneously over the mid-Atlantic, with a high probability of collision.

The air traffic control system which had evolved to handle this complex situation consisted of certain tracks across the North Atlantic along which pilots were directed. These tracks were separated by 120 (nautical) miles – since this distance was precisely 2 degrees of latitude and so ought to have avoided any confusion – and, along each track, planes were separated by 2000 feet in altitude. Thus there was a vertical separation of 2000 feet as well as a lateral separation of 120 miles. To ensure an adequate longitudinal separation, planes were assigned to a track at the same altitude only every quarter of an hour. Speed differences were not common in the 1960s, most aircraft flying at Mach 0.82. and so a simple time delay was adequate. Traffic flying in opposite directions was assigned to separate tracks at the same level, or separate levels on the same track, and one level was often reserved for crossing traffic.

As the total number of aircraft using this traffic control system grew, the position became intolerable for the airline operators. Aircraft were frequently required to accept routes 240 miles north or south of where they wanted to be or at the lower flight levels where fuel consumption increased enormously. Thus the airline owners approached the ICAO with a request that the lateral separation be lowered to 90 miles immediately and to 60 miles eventually. It was at this point that the confrontation broke out between the owners and the pilots; the pilots intuitively felt that these separations were unsafe and so refused to implement them. Thus a Systems Planning Group was set up by ICAO to consider the proposals; it officially consisted of only six members, but its meetings actually involved about one hundred delegates and observers with full protocal, and simultaneous translations into several languages.

2.3 Reliable data

During the early days of the discussions, the Federal Aviation Agency (FAA) had made a study of lateral errors while aircraft were in the Gander/Shanwick control area between Newfoundland and Ireland, and had concluded that a 90 mile minimum separation would be satisfactory. The pilots, however, were not impressed by the accuracy of this study – for several, technical, reasons – and so the FAA agreed to hold a hearing. The fight was on! The hearing finally endorsed the pilots' worries, and a new study was set up under the FAA and the Royal Aircraft Establishment at Farnborough. Thus, at the end of five years' continuous study, it was possible to make confident pronouncements about lateral errors! Some of the data are shown in Fig. 2.1; it is clear that planes over the North Atlantic spend about 1% of their time more than 40 miles off course laterally.

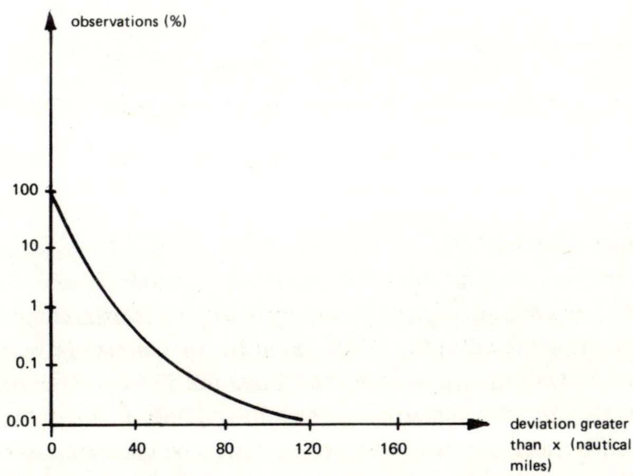

Fig. 2.1

Thus it was apparent that a 90 mile lateral separation was *not* adequately safe. Was there *any* solution which would satisfy both parties?

2.4 Optimisation

The natural reaction of many mathematicians when presented with a problem containing two competing alternatives is to seek an optimum solution. Here the two alternatives could be labelled 'cost of deviation' – the cost of extra fuel burned through not taking the shortest route, the loss of revenue carrying fuel instead of payload, the cost of extra staff on duty, and the like – and 'cost of collision' which (however one defines it) must decrease as the separation is increased. Thus the cost-benefit analysis would lead to a curve similar to that in Fig. 2.2, from which one could read off the minimum total cost. However, while the cost of deviation is merely tedious to work out (involving lots of well-understood but interlinked effects), the cost of collision involves putting a monetary value on a death – a difficult and politically unacceptable feat. Thus what seemed a perfectly acceptable mathematical approach came to grief for reasons of social acceptability – a warning to all modellers!

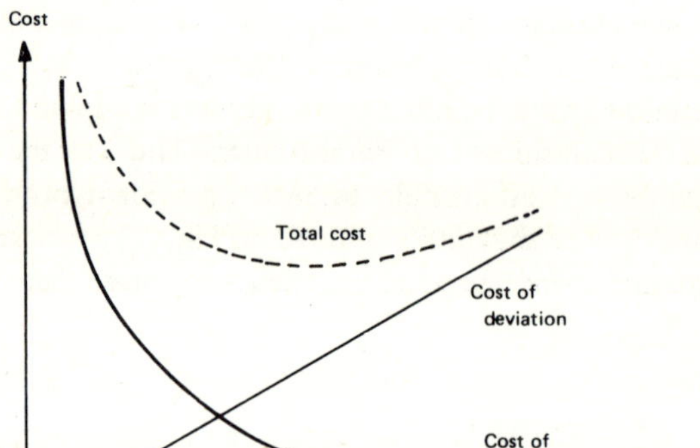

Fig. 2.2

2.5 Who needs a model?

If there are any readers still doubtful about *why* a model is necessary, let them consider the following question. If you are flying the North Atlantic and have a neighbour, at the same altitude, 120 miles north of you, is it safer if he is going in the same direction or in the opposite direction? The two arguments – which produce exactly opposite answers – go as follows. One argument says that same-direction traffic must be more dangerous, since the collision hazard exists for several hours (as opposed to only a few seconds when planes travelling in opposite directions pass). The other argument says that opposite-direction traffic must be more dangerous, since in this case it is certain that the longitudinal separation will be zero at some instant (whereas it may never be zero for same-direction traffic). At the end of the study it was possible to say that both carried the same risk: opposite direction traffic was about 30 times as dangerous, but there was 30 times less of it! How many readers would have predicted *this*?

2.6 Probabilities

Having failed with the optimisation approach the Systems Planning Group decided to evaluate the collision risk and compare it with a target level of safety. What value should have been taken for a target level, though? This was obviously a difficult and controversial subject, and the decision was not eased by the common acceptance that collisions were much more likely over a crowded airport area than over the North Atlantic. In the end, common sense came to the rescue. One collision per 10^6 flying hours was clearly too lax, since it would have involved about one collision every other year; one collision per 10^9 flying hours was too stringent, since the whole system would change dramatically in about ten years anyway. The number that was agreed eventually was about 0.5 collisions per 10^7 flying hours – something like ten times as safe as *any* flying in those days.

Now the probability of a collision for various separation distances had to be worked out. A typical aircraft considered by the Group (a Boeing 707) had a 150-foot wingspan, was 150 feet from nose to tail, and stood about 40 feet tall. Thus each of the two possibly colliding aircraft could be thought of as a $150 \times 150 \times 40$ feet box; if the two boxes touched this represented a collision. Easier to deal with is the equivalent situation of a $300 \times 300 \times 80$ feet box and a point; if the point entered the box this represented a collision. This model was clearly very rough and ready – aircraft could obviously pass closer together than this and not collide or, alternatively, could be effected by the wake from an aircraft some considerable distance ahead – but it provided a useful starting point. There is always plenty of time to add in details *after* one has understood the overall system!

The probability that the point would lie inside the box was the product of three separate probabilities – those of lateral, longitudinal, and vertical overlap. This, of course, assumed that the three probabilities were independent, and that planes were *not*, for instance, being assigned to adjacent tracks in such a way that they were side by side over most of the North Atlantic. Fortunately a direct test for this was available, since many planes had been timed going over the datum lines 20°W and 40°W. The data for proximate pairs were analysed and found to be uniformly distributed around a 15 minute mean (remember the quarter hour longitudinal separation?) – confirming the assumption of independence. Thus the total probability of a collision was given by the product of the probabilities in the three directions:

$$P(\text{collision}) = P_{\text{LAT}} \cdot P_{\text{LONG}} \cdot P_{\text{VERT}} \ .$$

Taking the easiest of these first, they considered P_{LONG}. Assuming the planes travelled at 480 knots (i.e. 8 miles per minute) with an average separation for planes on adjacent tracks of 25 minutes (from the data), then they were 200 miles apart. Taking the box length to be about 1/20 mile then gave P_{LONG} = 1/4000.

Next easiest was P_{VERT}, the probability of vertical overlap. Instrument design tended to ensure excellent vertical station keeping, and most of the times planes were oscillating about the assigned position with an amplitude of less than 100 feet. Thus P_{VERT} was quite high, about 1/4.

Finally came P_{LAT}, for which the statisticians came up with a value of 10^{-6} for the case of 90 mile separation. Multiplying all these probabilities together then gave

$$P(\text{collision}) = 0.625 \times 10^{-10} .$$

To convert this to an expected number of collisions per flying hour they needed to estimate how long it would take for the point to pass through the box. Estimating this from the relative velocities necessary to get sufficiently far off course in the time available gave a figure of about 3 seconds. Thus the collision risk was $0.625 \times 10^{-10}/3$ seconds, or 0.75 collisions every 10^7 flying hours.

As explained above, there remained many holes to be plugged in the model. The Systems Planning Group spent many months doing just that, and came up with the following amended figures: the collision risk per 10^7 flying hours was 0.6 with a 90 mile lateral separation and 0.1 with the original 120 mile separation. Thus 120 mile separation was 'safe' and 90 mile separation was 'unsafe'!

This is a gross simplification, of course; there was still a fair bit of disagreement about the actual probabilities, but the Group *was* agreed that 120 mile separation was about 6 times safer than 90 mile separation. How then to satisfy both the owners and the pilots? The answer came in the form of a suggestion from one of the pilots – to try staggered separation. Figure 2.3 shows the conventional separation pattern, with 120 mile lateral separation and 2000 feet vertical separation.

Since the vertical position-keeping was so good, the idea was to decrease the vertical separation while keeping the lateral separation the same. However, decreasing the separation from 2000 feet to 1000 feet was too much, and so a staggered pattern (as shown in Fig. 2.4) was suggested.

The detailed data and analyses mentioned earlier allowed the various collision risks for the new configuration to be worked out accurately. It is fairly easy to see, however, that the risk ought to be about the same as for the standard configuration: the probability of collision with diagonal neighbours has been increased, but this has been compensated for by the edge effect (where the aircraft at the edges of the pattern have fewer neighbours). In fact, the total collision risk was slightly reduced by the staggered separation pattern!

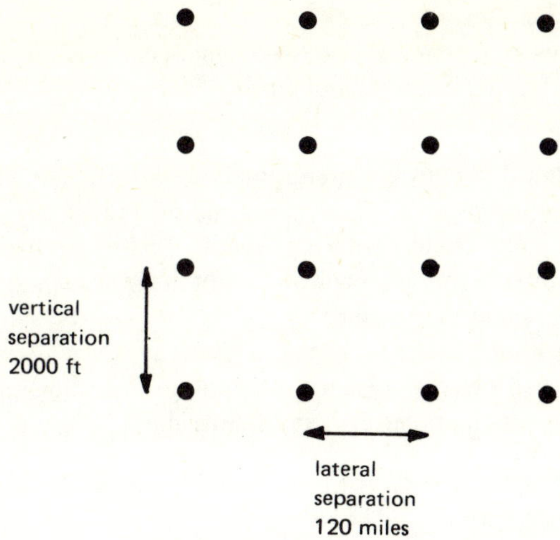

Fig. 2.3

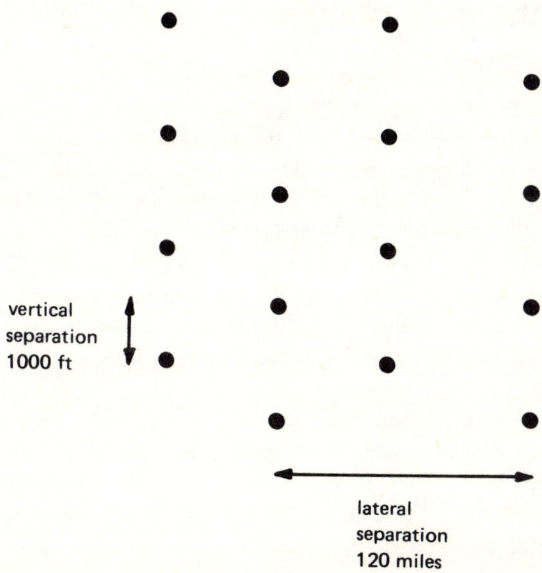

Fig. 2.4

The advantages for all concerned are quite amazing. First, of course, is that it was safer than the conventional stagger pattern *for the pilots.* All clearances can be given in whole numbers of degrees (being integer multiples of 60 miles) and in multiples of 1000 feet – *good for the controllers.* Finally, the width of the whole flight pattern is reduced from the original 360 miles to only 180 miles – a greater reduction than that achieved by the 90 mile separation plan, and therefore more economical *for the operators.*

2.7 Implementing the changes

Even though a solution was now available which would seem to satisfy all parties, the Group's job was not over yet. As Machol puts it:

> "Having convinced ourselves of the advantages of staggered separation, we had to be very careful in convincing everybody else. In the first place, the pilots had been supporting us all those years while we fought for 120 miles instead of 90, and there could have been trouble if word got back to them that we were bought off and accepted 60 . . . In the second place, the controllers were not sure that they could set up such a staggered system without additional complications, but they agreed to give it a try when we agreed to drop the request for extra segregation of opposite direction traffic . . . Finally, it had to be tested, monitored, and followed up."

2.8 What to learn from the story

It was possible to solve the problem for two main reasons: understanding was developed of the system under consideration, and wisdom was exercised in the political arena.

At the beginning of the discussions little hard information was available about near misses and the like. The initial survey was carried out by the (very reputable) FAA, but the gut reaction of the experienced pilots was that the scientists had made a mistake somewhere. Further investigations – a very time-consuming business – led to reliable data, on which all the later calculations were based, and an understanding of the whole system.

The problem, however, did not exist merely in a neat, mathematical world. The optimisation approach, for instance, broke down not because of the mathematics but becuase it was socially and politically unacceptable to put a costing on a fatality. One must be very careful to include *all* the important effects when analysing a system.

The final point worth making, when looking back over the story, is that the eventual answer was not the product of some sophisticated mathematical or computing technique but rather a commonsense solution. The minimum of mathematics was used; the emphasis was rather on getting the facts right and understanding the real problems.

3 WHALING

3.1 Introduction

As technology becomes more sophisticated, man's capacity to exploit his environment becomes greater. In many respects this is an asset because he can make areas of the earth productive which, a few years ago, were thought to be of no value. This applies both to agriculture and to natural resources such as oil, gas, and coal. Set against these advantages there have, of course, been disadvantages. Amongst these have been the detrimental effects on many animal populations. Ironically, because of the improvements in communication brought about by modern technology, many of us are much more aware of environmental problems than we would have been say twenty years ago. Conservation groups of various kinds are very active now, and their activities have been well publicised by the media.

One area of environmental exploitation which is fairly frequently brought to our attention is whaling. It is reckoned that some species of whale are being brought to the edge of extinction by over-exploitation, and the activities of one conservation group – the Greenpeace organisation – have been reported frequently on television. An annual event which is brought to our attention is the meeting of the International Whaling Commission (IWC) at which whaling quotas are set for the next year.

Although the problems of natural resource management have only relatively recently come to the attention of the general public, scientists have been aware of them for many years. For whaling, in particular, there is an impressive quantity of statistics which dates from the formation of the Bureau of International Whaling Statistics in Norway in 1930. Although certain controls on whaling were being imposed as long ago as the 1930s, there has been a reluctance by whalers and governments to accept that whale conservation is a problem. This attitude caused J. E. Hamilton to write in *Nature*†, as long ago as 1948:

> "In the presence of vast and unestimated number of animals which it is desired to exploit commercially, nothing is easier than to persuade oneself that the stock is so great that it is impossible to injure it, still less to destroy it."

Figure 3.1 shows the decline of the number of sei whales in the western Indian Ocean, so there has obviously been little attention paid to the warnings of scientists such as Hamilton.

We have included a discussion of this problem because, quite unexpectedly, mathematics does have an important part to play. Mathematical models are used to predict the effect of various harvesting policies on different populations of whales, and these predictions are available for use by decision-makers in the formulation of their policies. We should not be naïve enough to expect that mathematics will provide definitive answers in a politically sensitive area such as this. What we will see, however, is that mathematics gives us information which could never have been arrived at by intuition.

At a macroscopic level, the life history of whales is similar to that of other mammals. That is, they are born, they mature and produce offspring, and

† Hamilton, J. E. (1948), Effect of present-day whaling on the stock of whales, *Nature*, **161**, 12 June, 913–914.

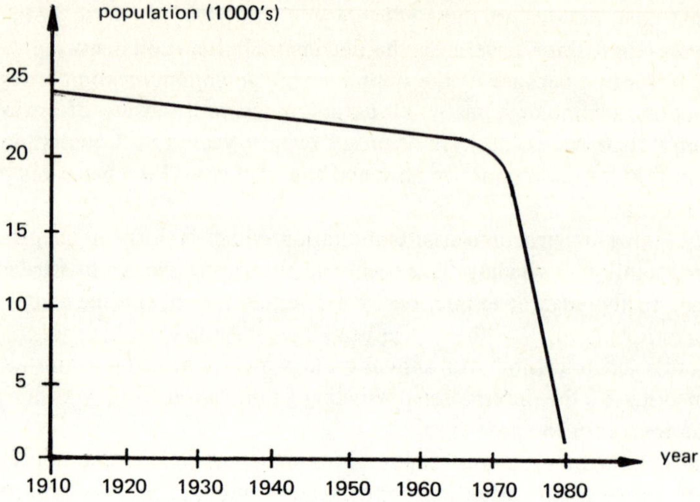

Fig. 3.1

eventually die. An additional aspect in this case is that the whales are harvested. Any attempt to model the dynamics of whale populations must take these basic facts into account. A simple approach to population modelling is to treat the population growth as a continuous process, which can be a reasonable approximation when dealing with large populations.

3.2 Analysis

Denoting the population at time t by $N(t)$ we can write

$$dN/dt = [r-(M+F)]\ N \quad . \tag{3.1}$$

where r is the recruitment factor (measured in recruits/head/year) and indicates the rate at which the population would be increasing with no other deaths or losses through harvesting, whilst M is the natural mortality factor (deaths/head/year) and F is the fishing mortality factor (number caught/head/year).

The solution of model (3.1) is

$$N(t) = N_0 \exp\ [r-(M+F)]\ t \ ,$$

where N_0 is the stock level when $t = 0$.

If the population is represented by model (3.1), it will have the following characteristics.

(i) It will be in equilibrium when $r = M+F$.
(ii) It will grow (exponentially) if $r > M+F$.
(iii) It will decay (exponentially) if $r < M+F$.

Clearly, then, if there is over-fishing the population could be reduced very rapidly. The model, therefore, could be used, as it stands, to predict the outcomes of various levels of fishing effort. That is, knowing values of r, M, and the population level at a particular time, the population remaining in a known number of years could be predicted for various values of F. It is obviously important to have good estimates of r, M, and the population level. These are not easy to obtain because whales are not easily observed directly. It is now known that errors in parameter estimation have themselves caused over-exploitation of certain whale populations. For example, one of the methods of estimating population levels has been to relate them to what is called *catch per unit effort*. The basis for this is simple – if the population is large then more whales will be caught for the same amount of effort than if the population is small. Until fairly recently it was assumed that population (or 'stock level') was directly proportional to catch per unit effort (heavy line in Fig. 3.2). As whaling technology has improved over the years it has been difficult to identify a standard measure of effort, but one has been devised and seems to have produced reasonable results. What was overlooked was that, although whales are more easily spotted when populations are high, there is an upper limit to the number of whales which can be handled by a team in a day. The catch/unit effort with stock level relationship, therefore, is probably closer to the dotted line in Fig. 3.2. Some thought will show that this error has led to underestimates of the decline in stock levels.

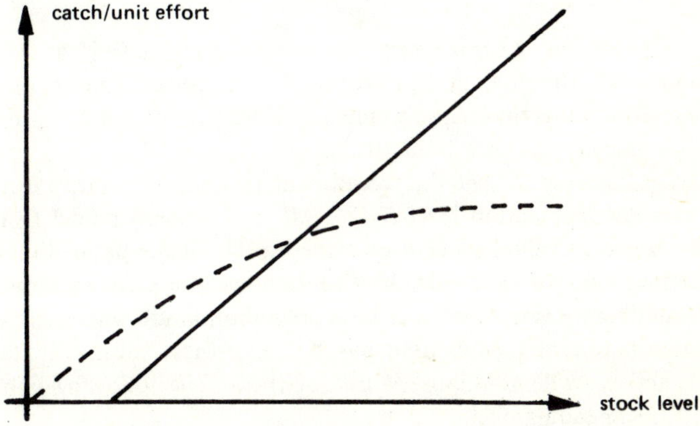

Fig. 3.2

Estimates of stock levels play an important role in management policies. Another important concept used in policy making is that of maximum sustainable yield (MSY). Suppose we consider a population of whales which has not been exploited by man. Intuition tells us that if we start removing some of the population this will, at first, improve the environment for the remainder, and the population will grow more rapidly to offset the numbers harvested. The yield from the population will thus increase until it reaches a maximum – the MSY. Harvesting beyond this level, however, means that the population is being depleted faster than it can reproduce itself. Figure 3.3 illustrates this intuitive picture.

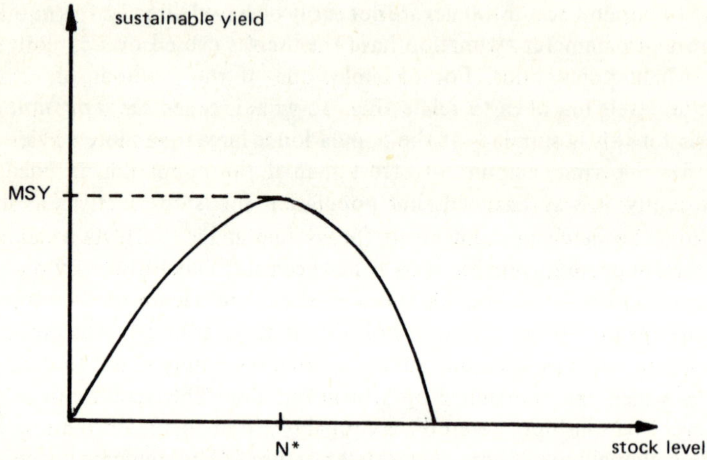

Fig. 3.3 Relationship between population level and maximum sustainable yield.

A way of making use of this concept is to set the target yield at 90% of MSY, and couple with this the total protection of the species if the population falls more than 10% below that corresponding to MSY (i.e. 90% of N^* in Fig. 3.3).

Before moving on, we note that as far as model (3.1) is concerned, the above discussion reveals that the recruitment rate r and mortality factor M both depend on the population level N. We will not develop model (3.1) further, however, because it does gloss over some aspects of the population dynamics. For example, whales do not reproduce continuously but have an annual breeding season. Another relevant factor is that only the mature whales are capable of reproduction. Recent models used by IWC take these additional features into account, and we will now look at one of these. The following is based on a paper by J. R. Beddington.†

A class of discrete models for whale harvesting takes the form of a time-delayed nonlinear difference equation

$$N_{t+1} = F_1(N_t) + F_2(N_{t-k}) - C_t \ , \tag{3.2}$$

† J. R. Beddington (1978), On the risks associated with different harvesting strategies, *Rep. Int. Whal. Commn.*, **28**, 165.

where N_t denotes the stock level in year t,
$F_1(N_t)$ is the number of whales surviving from year t to year $t+1$,
$F_2(N_{t-k})$ is the recruitment to the stock from the breeding population k years previously,
and C_t is the harvesting quota.

Such a system would be in equilibrium when $N_{t-k} = N_t = N_{t+1} = N^*$, and the corresponding catch quota would be given by

$$C_t = F_1(N^*) + F_2(N^*) - N^* \ . \tag{3.3}$$

This equation reveals that the catch quota depends on the equilibrium population, N^*, and by differentiating (3.3) with respect to N^*, the condition for MSY can be obtained as

$$F_1'(N^*) + F_2'(N^*) - 1 = 0 \ . \tag{3.4}$$

In practice, as we have seen already, it is common to harvest at a level less than MSY. This level is usually defined as a percentage of MSY – for example 90% of MSY. In doing this, C_t is fixed and the equilibrium stock level will be different from the stock level corresponding to MSY. The resulting equilibrium stock level is obtained by solving the equation

$$C_t = F_1(N^*) + F_2(N^*) - N^* \ , \tag{3.5}$$

in which, this time, C_t is fixed. For most forms of F_1 and F_2, equation (3.5) will have two roots – corresponding to two possible equilibrium stock levels. This situation is illustrated in Fig. 3.4.

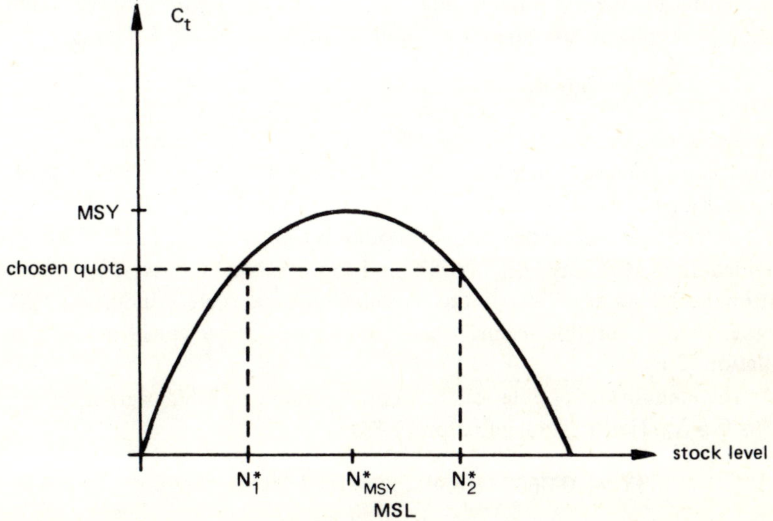

Fig. 3.4

Rather than proceed further at this general level, however, let us look at a specific model. Beddington quotes the following model as having been used in 1976 in the assessment of quotas for sei whales:

$$N_{t+1} = 0.94\,N_t + N_{t-8}\left[0.06 + 0.0567\left\{1 - \left(\frac{N_{t-8}}{N_0}\right)^{2.39}\right\}\right] - 0.94\,C_t \tag{3.6}$$

where N_0 is the unexploited stock level.

Working from first principles, we note that the equilibrium stock level is achieved when

$$N_{t+1} = N_t = N_{t-8} = N^* \ .$$

Using this in equation (3.6) leads quickly to

$$0.0567\,N^*\left\{1 - \left(\frac{N^*}{N_0}\right)^{2.39}\right\} - 0.94\,C_t = 0 \tag{3.7}$$

as a condition for equilibrium.

From equation (3.7) we note that the harvest quota, C_t, is related to the equilibrium population level by

$$C_t = 0.0603\,N^*\left\{1 - \left(\frac{N^*}{N_0}\right)^{2.39}\right\} \ . \tag{3.8}$$

Differentiating the right-hand side of equation (3.8) with respect to N^* and setting the result to zero gives the condition for C_t to be MSY. This is

$$N^* = 0.6\,N_0 \ .$$

In plain words, this means that the MSY will be achieved when the whale population has been reduced to 60% of the unexploited level. This, in itself, is an interesting and useful result, but there is much more to come.

As we saw earlier, the common policy is to harvest not at MSY but at some level less than MSY, say 90% of MSY, and to put a ban on harvesting when stocks fall below let us say 90% of the equilibrium level corresponding to MSY. Let us see, then, what this means when applied to the sei whale model given by equation (3.6).

The equilibrium population, N^*, corresponding to MSY is given by $N^* = 0.6\,N_0$. Hence, from equation (3.8),

$$\begin{aligned} \text{MSY} &= 0.0603\,(0.6\,N_0)\,(1 - 0.6^{2.39}) \\ &= 0.02551\,N_0 \ . \end{aligned} \tag{3.8a}$$

If the actual harvest quota, C_t, is a fraction of MSY,

$$C_t = P.\,\text{MSY} \ ,$$

then we can find the corresponding equilibrium population, N^*, from

$$0.02551\ .P.\,N_0 = 0.0603\,N^* \left\{1 - \left(\frac{N^*}{N_0}\right)^{2.39}\right\} .$$

Denoting N^*/N_0 by M we can write this as

$$0.02551{\cdot}P = 0.0603\,M \left\{1 - M^{2.39}\right\} ,$$

or alternatively,

$$M - M^{3.39} - 0.42289\,P = 0 \ . \qquad (3.9)$$

Given a value of P, equation (3.9) can be solved by numerical methods – the Newton Raphson method will do – and gives two roots in each case. Table 3.1 gives these roots for a range of values of P and gives the corresponding yield – which is the same for both equilibrium levels. Figure 3.5 illustrates these results graphically.

Table 3.1 – Effect of harvest quotas on equilibrium stock levels.

P	$\frac{N^*}{N_0}$ lower	$\frac{N^*}{N_0}$ upper	$\frac{\text{Yield}}{N_0}$
0.9	0.445	0.737	0.023
0.8	0.374	0.792	0.020
0.7	0.316	0.832	0.018
0.6	0.265	0.865	0.015
0.5	0.217	0.893	0.012
0.4	0.172	0.918	0.010
0.3	0.128	0.941	0.008
0.2	0.085	0.962	0.005
0.1	0.042	0.982	0.003

Here the model has yielded information which certainly could not have been obtained by intuitive means. Who would have guessed that there could be two equilibrium stock levels for each harvest quota level? However, what do these two equilibrium population levels really mean?

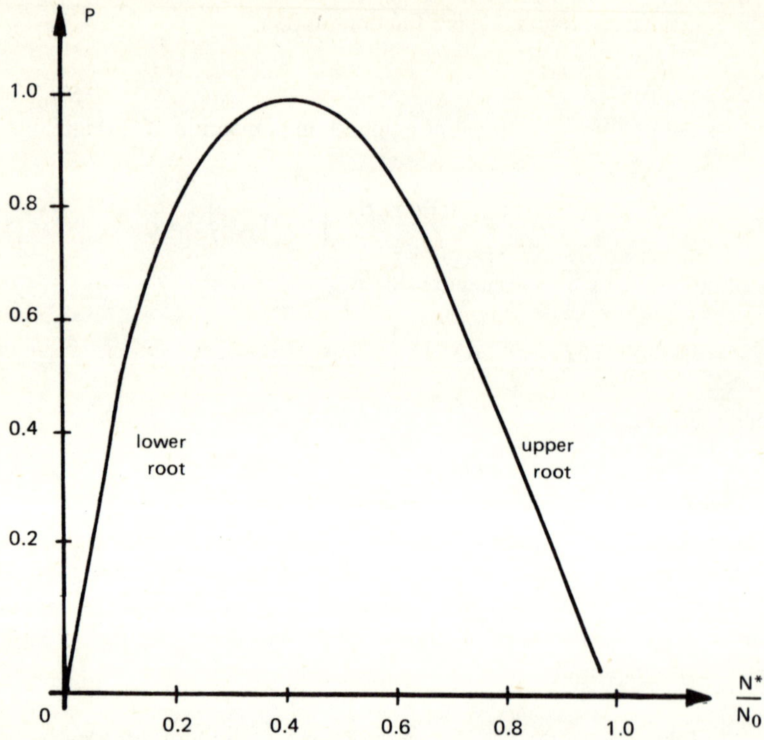

Fig. 3.5

Mathematicians always look closely at equilibrium points in dynamic systems, because experience has taught them that strange things can happen there. Most readers will be familiar with the two equilibrium states of a cone. If we place a cone on its base (Fig. 3.6a) then it will be in equilibrium. Furthermore , if it is rocked slightly (Fig. 3.6b) it will return to its equilibrium position. The equilibrium position is then said to be STABLE. If, on the other hand, it is balanced very carefully on its apex (Fig. 3.6c) then it will rest in equilibrium, but if it is disturbed (Fig. 3.6d) even very slightly, it will fall over. The cone, in this case, is in UNSTABLE equilibrium.

With this analogy in mind, therefore, it would seem sensible to investigate the equilibrium states of the whale population. We have to ask ourselves the following question. 'If the system is slightly disturbed from its equilibrium state, will it return to equilibrium (stable situation) or will the disturbance be amplified (unstable situation)?' This question can be answered by carrying out a stability analysis. Such an analysis is rather messy, but shows that for each of the pairs of equilibrium points in Fig. 3.5, the lower point is unstable and the upper stable.

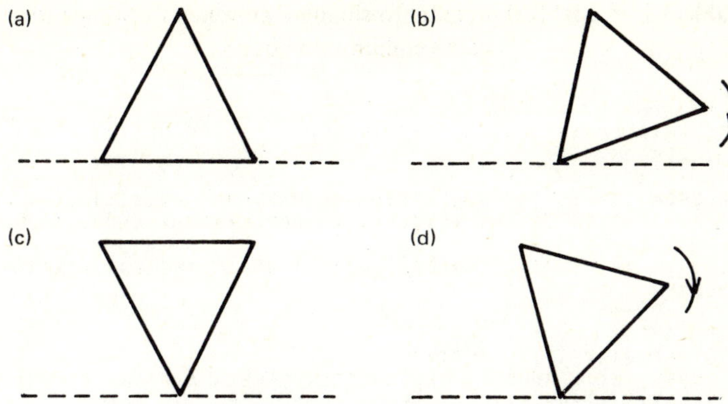

Fig. 3.6 – Equilibrium states of a cone.

It can also be shown that the equilibrium point corresponding to MSY has a kind of stability.

Another related question is the following. 'If the population is in a given state and a policy of harvesting at a given percentage of MSY is introduced, will the population reach equilibrium, and if so at which equilibrium point?' Readers with access to a computer – even a small microcomputer – can use the simple program in Table 3.2 to investigate what happens. The program statements are scaled so that it works in units of N_0. That is, the unexploited population would be 1.

For example, in Fig. 3.1 we can see that in the years between 1940 and 1960, the sei whale population in the western Indian ocean was constant – at about 80% – 85% of 1900 population. If we take the 1900 population as a good approximation to the unexploited population, then we could supply 0.8 as the value of S in the program. Supplying 0.9 as the value of P, we can see that the population eventually reaches equilibrium at the value of 0.737 predicted by equation (3.9) and shown in the first line of Table 3.1.

Some experimentation with other initial population structures and values of P will confirm that the lower equilibrium points in Table 3.1 are unstable and the upper points are stable. Another interesting feature is that the lower the harvest quota, i.e. value of P, the more quickly the system reaches the stable equilibrium point. This reveals that the nearer the harvest quota is to MSY, the less stable the system is in a sense. This means that in using a policy based on harvesting very near MSY, the scientists involved have to be very careful about getting the parameters of the model correct.

Table 3.2 – A BASIC program to simulate movement of population to stable equilibrium points.

```
REM THIS PROGRAM SIMULATES THE MOVEMENT OF A POPULATION TO A STABLE
REM EQUILIBRIUM
REM
REM THE USER INPUTS THE PROPORTION OF MAXIMUM SUSTAINABLE YIELD (MSY)
REM AND THE INITIAL STOCK LEVEL (I.E. POPULATION) IN UNITS OF NO
REM
REM THE YEAR AND STOCK LEVEL ARE OUTPUT UNTIL THE EQUILIBRIUM IS
REM ACHIEVED
REM
REM T IS TIME
REM C IS THE HARVESTING QUOTA
REM THE ARRAY N(I) HOLDS 9 YEARS STOCK LEVEL, N(9) BEING THE LATEST
REM N1 IS NEXT YEARS STOCK LEVEL
REM
10 LET T=0
20 PRINT 'INPUT PROPORTION OF MSY'
30 INPUT P
40 PRINT 'INPUT INITIAL STOCK LEVEL AS A PROPORTION OF NO'
50 INPUT S
60 FOR I = 1 TO 9
70 LET N(I) = S
80 NEXT I
90 LET C=0.94*P*0.02551
100 PRINT 'YEAR','STOCK LEVEL'
110 PRINT '----','-----------'
120 PRINT T,S
130 LET T=T+1
140 LET N1=0.94*N(9)+N(1)*(0.06+0.0567*(1-N(1)^2.39))-C
150 PRINT T,N1
155 IF (N1-N(9))/N1 < 1E-4 GOTO 210
160 FOR I = 1 TO 8
170 LET N(I)=N(I+1)
180 NEXT I
190 LET N(9)=N1
200 GOTO 130
210 END
```

3.3 Review

What we have seen in this chapter is an application of mathematics to a very topical and politically controversial situation. The interesting thing about this particular application is that it reveals some features of the behaviour of the system which could never have been achieved by intuitive methods or by experiments. These features were:

(i) the existence of two equilibrium points – one stable the other unstable,
(ii) the slower approach to equilibrium with harvest levels close to MSY,
(iii) the real chance of damaging the population through not getting accurate values for the parameters of the model.

Part II

Mathematical Modelling

INTRODUCTION TO PART II

In this section we are expecting you to work harder. There are 20 case studies here which have clear problem statements, and we very much encourage you to attempt to find a solution to these real problems before looking at the solutions that we have given. You might well use different sorts of mathematics from ours, but you might also achieve better solutions than the ones we have given. Remember that we are trying to solve real problems, and unlike the usual mathematical analysis that you have probably had before, there is no unique solution to each of these problems. In fact, it is often quite difficult to know whether the solution that you have obtained is a good one. On the other hand, do not feel that you must attempt every problem here. A great variety of topics has been selected, and we strongly suggest that you choose only the ones that really interest you. You will find that wanting to know solutions to problems is a great motivating factor for actually getting down and doing just that. Again, we hope you enjoy these case studies and hope that, as in the first section, you will begin to see the way in which mathematics is applied to real problems. It is not usually the mathematical analysis which is the most difficult, but in many problems the formulation stage is often the most taxing part.

1 HANDICAPPING WEIGHTLIFTERS

1.1 Problem

Weightlifting is a minority sport, but it is a sport which can be understood readily by the layman. There are nine official bodyweight classes (see Table 1.1) and there are two principal lifts, namely the *snatch* and the *jerk*. In the snatch, the weight is pulled from the floor to a locked arm overhead position in a *single* move, although the lifter is allowed to move or squat under the weight as it is being lifted, whereas for the jerk, *two* movements are allowed, the first to the chest and the second to the overhead position.

Table 1.1 lists the world weightlifting records, current on 31 July 1977. The maximum bodyweight for each class is also shown.

Table 1.1

Bodyweight Class	(kg)	Lift	Lifted (kg)	Name and Country	Date
Flyweight	(52)	S	109	A. Voronin (USSR)	18 Mar 1977
		J	141	A. Voronin (USSR)	18 Jul 1976
Bantamweight	(56)	S	120.5	K. Miki (Japan)	25 Oct 1976
		J	151	M. Nassiri (Iran)	2 Aug 1973
Featherweight	(60)	S	130	G. Todorov (Bulgaria)	25 May 1976
		J	161.5	N. Kolesnikov (USSR)	20 Jul 1976
Lightweight	(67.5)	S	141.5	A. Aibazian (USSR)	15 Jul 1977
		J	180	S. Pevzner (USSR)	16 Jul 1977
Middleweight	(75)	S	157.5	Y. Vardanyan (USSR)	7 May 1977
		J	195	V. Milotosyan (USSR)	30 Jan 1977
Light-Heavyweight	(82.5)	S	170	B. Blagoyev (Bulgaria)	25 May 1976
		J	207.5	R. Milser (W. Germany)	8 Apr 1976
Middle-Heavyweight	(90)	S	180	D. Rigert (USSR)	14 May 1976
		J	221	D. Rigert (USSR)	14 May 1976
Heavyweight	(110)	S	185	V. Khristov (Bulgaria)	10 Apr 1976
		J	237.5	V. Khristov (Bulgaria)	22 Sep 1975
Super-Heavyweight	(over 110)	S	200	K. Plachkov (Bulgaria)	25 May 1976
		J	255	V. Alexeev (USSR)	27 Jul 1976

S – snatch, J – jerk

Table 1.2

Bodyweight Class	Total Winning Lifts		
	S	J	TOTAL
Flyweight	105	137½	242½
Bantamweight	117½	145	262½
Featherweight	125	160	285
Lightweight	135	172½	307½
Middleweight	145	190	335
Light-heavyweight	162½	202½	365
Middle-heavyweight	170	212½	382½
Heavyweight	175	225	400

It is clear that as bodyweight increases so does the lift, but what is not clear is the functional form of this increase. The problem is not just a physics/mathematics problem, but is a real practical one, as for many competitions (e.g. school, college) there are just not sufficient numbers to warrant the utilisation of the nine weight classes. It is also of some interest, even in serious competitions, to have an 'overall' winner. So the problem is to design some form of handicapping, which compensates for the various bodyweights.

Having formulated a method of handicapping weightlifters, use it to find an *overall* winner for the winning snatch lifts at the 1976 Montreal Olympic Games, which are detailed in Table 1.2.

1.2 Mathematical models for weightlifting

There have been many models formulated which have been used for handicapping weightlifters. We will firstly describe several of these models, and then in section 1.3 use them to answer the question (and compare the results).

(i) *Linear model*

In Fig. 1.1, we have illustrated the world record data points from Table 1.1 for both the snatch and the jerk, assuming that lifters are at their maximum allowed weight for their class. It is evident that a straight line is not a very accurate model, but nevertheless it does have some use. For example, the television programme "TV Superstars" uses the model

$$L' = L - B \ ,$$

where L' is the handicapped lift, L the lift, and B the lifter's bodyweight. This model implies a linear relationship between L and B with a unit slope, and a check with the data points in Fig. 1.1 indicates that this approximation is not very accurate. At the level of performance of superstars it turns out to be much worse, favouring the lighter competitors substantially. On the other hand, the model has as its virtue *simplicity.* It can be understood by the average television viewer, and for this purpose can be viewed as a 'good' model. Entertainment is the object of TV Superstars, and a more complicated form of handicapping might leave viewers in some doubt as to what was happening.

(ii) *Power law models*

Looking at the data points in Fig. 1.1 suggests that some power law model should be a more accurate description than a linear model. So if we assume that $L = k\,B^{\alpha}$ where k and α are constants, then

$$\log L = \log k + \alpha \log B \ ;$$

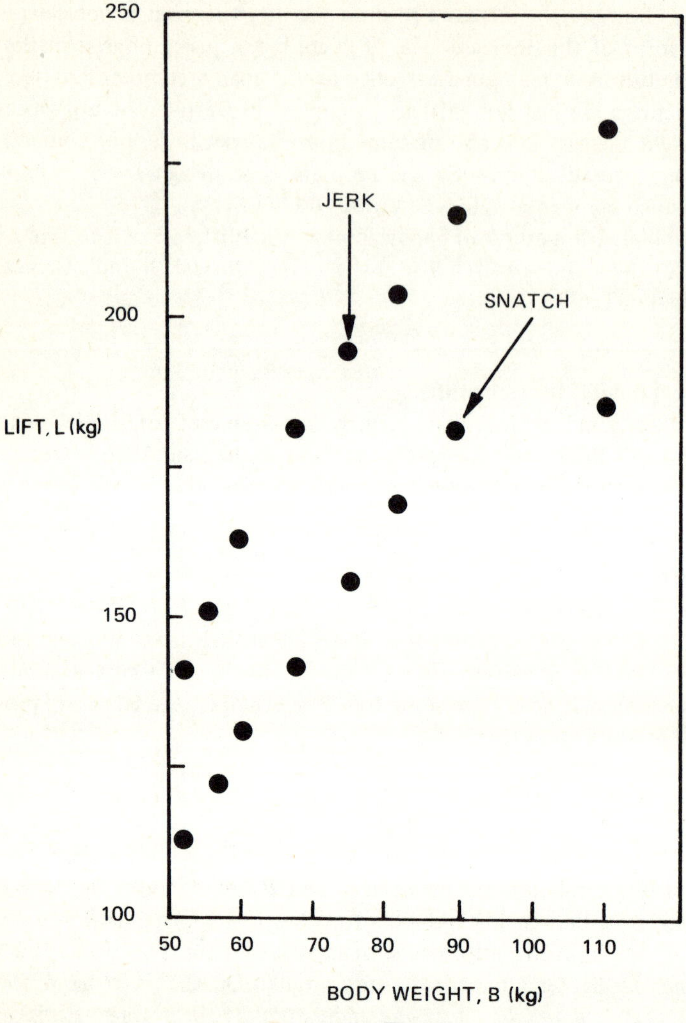

Fig. 1.1

this means that a log L – log B graph should be a straight line, with slope α. For the world record data given above we obtain the points illustrated in Fig. 1.2. A straight line looks a reasonable approximation for all the data points except the heavyweight. Remembering, however, that the super-heavyweight data have not been included, a power law model does not seem entirely satisfactory, particularly for the upper weight classes. In fact, two power laws have been used extensively. These are as follows.

(a) *Austin formula* which is a 3/4 power law,

i.e. $$L' = L/B^{3/4} \tag{1.2}$$

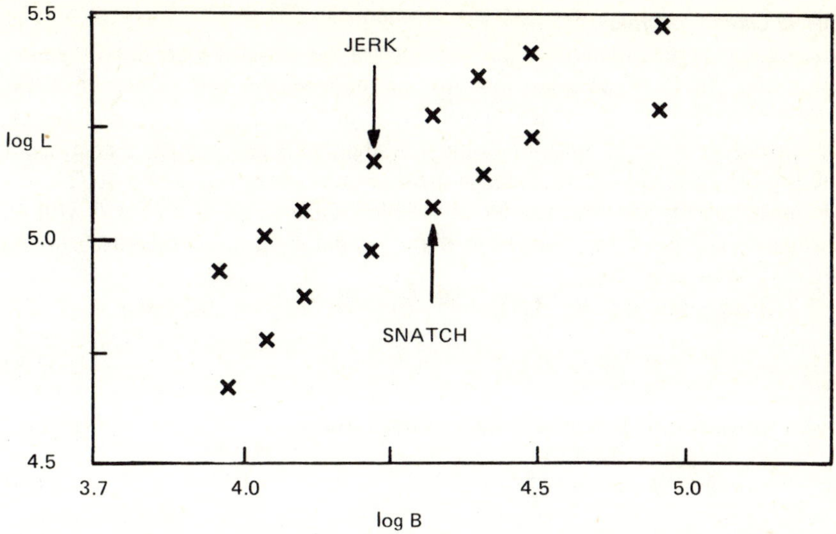

Fig. 1.2

and was used some 20 years ago in the UK for odd lift comparison and championship best awards

(b) *Classical formula* which is a 2/3 power law,

i.e. $$L' = L/B^{2/3} \tag{1.3}$$

and has been used for some time. This model can in fact be deduced from physiological arguments in simple terms. Briefly we assume that the lift is proportional to the average cross-sectional area of the lifter's muscle, say A,

i.e. $$L = k_1 A \; . \tag{1.4}$$

We also assume that A is proportional to the square of a typical body length, say l,

i.e. $$A = k_2 l^2 \; , \tag{1.5}$$

and finally, the 'typical' body length is related to the bodyweight by

$$B = k_3 l^3 \; . \tag{1.6}$$

With these assumptions,

$$L = k_1 k_2 \, l^2 = k_1 k_2 (B/k_3)^{2/3} = k \, B^{2/3} \; . \tag{1.7}$$

(iii) *O'Carroll formula*

The arguments presented above for the classical 2/3 power law are not altogether convincing. In 1967, an improved formula, designed by O'Carroll (1967) was introduced, and it has since gained wide acceptance by the weightlifting authorities. It is based both on zoological arguments and statistical analysis of the top 50 worldwide lifters for each division, over all time up to 1964. This was the most 'robust' and broadest based sample available, and it was used with an automatic compensation for the differing populations of competitors in the various bodyweight classes.

The assumptions (1.4), (1.5), and (1.6) above are generalised to:

$$L = K_1 A^x, \; x < 1 \; , \tag{1.8}$$

which allows for the loss of efficiency at large size;

$$A = K_2 l^y, \; y < 2 \; , \tag{1.9}$$

which allows for variations in shape. Here l is now a muscular length scale and

$$B - B_0 = K_3 l^3 \; . \tag{1.10}$$

This treats the total bodyweight $B = B_0 + B_1$ as the sum of B_0, being non-muscular weight such as cerebral which is almost independent of size, and $B_1 = K_3 l^3$ denoting the varying muscular and related weights. Now from (1.8), (1.9), and (1.10) we deduce a formula of the form

$$L = K(B - B_0)^\beta \; ,$$

where K and β are constants. From the statistical analysis in O'Carroll (1967), the values of the parameters are taken as

$$B_0 = 35 \text{ kg}, \;\; \beta = 1/3 \; ,$$

giving the model as

$$L = K(B - 35)^{1/3} \; . \tag{1.11}$$

Using this model, the handicapped lift is

$$L' = \frac{L}{(B - 35)^{1/3}} \; . \tag{1.12}$$

In Fig. 1.3 we illustrate a plot of log L against log $(B - 35)$ for the data given in the introduction. We also illustrate straight lines with slope 1/3, which indicate the accuracy of this model. The three best performing classes (middle, light-heavy, and mid-heavy) are relatively highly populated with competitors, whereas the extreme bodyweight classes have relatively few competitors. This is probably the major factor behind the differences between the population-compensated formula and the population-affected world record figures.

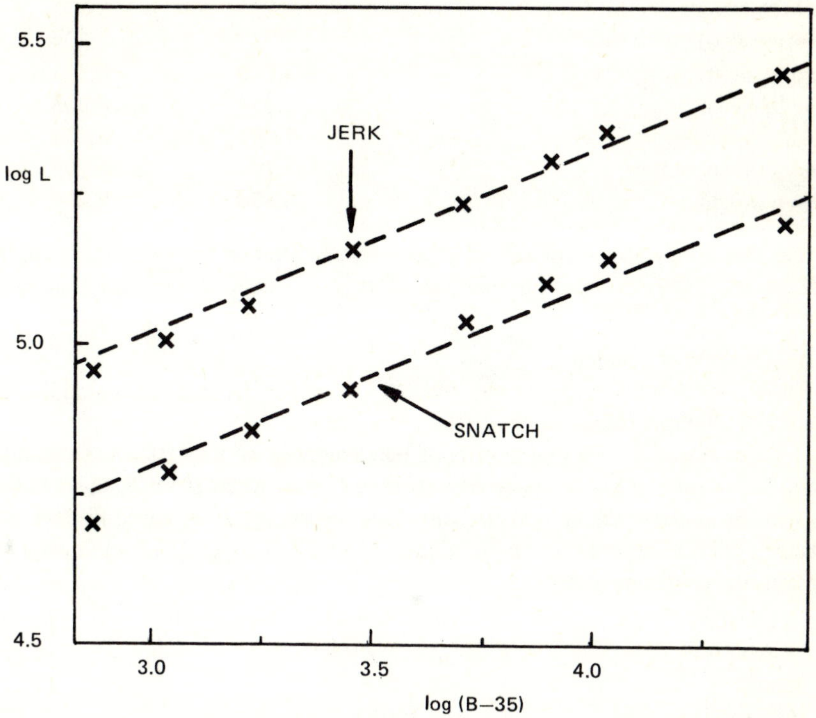

Fig. 1.3

(iv) *Vorobyev formula*
This is a Russian formula which is based on determining a 'degree of merit' index n, which is given by

$$n = \frac{L + B}{B\,[0.45 - (B - 60)/900]} \ . \tag{1.13}$$

Note that $L + B$ is based on the idea that the lifter lifts himself along with the barbell, so that the factor $(L + B)/B$ is the total lift as a multiple of bodyweight. The factor $[0.45 - (B - 60)/900]$ represents the proportion of available usable weight, which decreases with body weight.

The degrees of merit values for the world record lifts are given in Table 1.3.

Table 1.3

Bodyweight class	Degrees of merit Snatch	Jerk
Flyweight	6.747	8.088
Bantamweight	6.936	8.135
Featherweight	7.037	8.204
Lightweight	7.010	8.301
Middleweight	7.154	8.308
Light-Heavyweight	7.201	8.371
Middle-Heavyweight	7.199	8.293
Heavyweight	6.800	8.010

Using this handicapping model, the best snatch lifter is the light-heavyweight (Blagoyev of Bulgaria) and the best jerk lifter is the middleweight (Milotosyan of the USSR).

1.3 Solutions of problem

We now use the models to find the 'overall' winner for the snatch winning lifts in the 1976 Olympic Games.

To compare the various models of handicapping we must select a common base; we adapt the handicapped lifts so that $L' = L$ when $B = 75$, the middleweight lifter's maximum bodyweight. That is to say, L' is an equivalent lift made at middleweight. From equations (1.1), (1.2), (1.3), (1.12), and (1.13). we obtain the handicapped lifts

1. Superstars $\quad L' = L - B + 75 \quad$ (1.15)
2. Austin $\quad L' = L\,(75/B)^{3/4} \quad$ (1.16)
3. Classical $\quad L' = L\,(75/B)^{2/3} \quad$ (1.17)
4. O'Carroll $\quad L' = L\,[40/(B - 35)]^{1/3} \quad$ (1.18)
5. Vorobyev $\quad L' = [29250\,(L + B)\,B\,(465 - B)] - 75 \quad$ (1.19)

The handicapped lifts are given below for their *snatch* lift together with their numerical positions (assuming each lifter is at his maximum allowed bodyweight).

Bodyweight class	Snatch lift	HANDICAPPED LIFTS Superstars		Austin		Classical		O'Carroll		Vorobyev	
Flyweight	105	128.0	8	138.2	7	134.0	8	139.7	8	138.8	7
Bantamweight	117.5	136.5	7	146.3	4	142.8	6	145.7	4	146.6	4
Featherweight	125	140.0	5	147.7	3	145.0	3	146.2	3	147.7	3
Lightweight	135	142.5	4	146.1	5	144.8	5	144.7	6	145.8	5
Middleweight	145	145.0	3	145.0	6	145.0	3	145.0	5	145.0	6
Light-heavyweight	162.5	147.7	2	151.3	1	152.5	1	153.5	1	152.1	1
Middle-heavyweight	170	155.0	1	148.3	2	150.5	2	152.9	2	150.3	2
Heavyweight	175	140.0	5	131.3	8	135.6	7	141.9	7	138.5	8

Perhaps what is most surprising about these handicapped lifts is their agreement. With the exception of the Superstars model, all the models give the same first three places. In fact the Austin and Vorobyev models have complete agreement in all the rankings, and for all but the heavyweight the actual values of L' are very close. So, on the evidence here, we conclude that the best snatcher among the Montreal Olympics champions was the light-heavyweight, V. Shery (USA).

1.4 Reference

O'Carroll, M. J., (1967), On the relation between strength and bodyweight, *Research in Physical Education*, **1**, 6.

1.5 Related Problems

(1) Design a handicapping system for (i) running races, (ii) swimming races, so that competitors can be timed over their preferred distances and compared.

World records for running

MEN

Distances	*Time* *hr*	*min*	*sec*	*Name*	*Nation*	*Year*
100 m			9.95	J. Hines	USA	1968
200 m (turn)			19.72	P. Mennea	Italy	1979
400 m			43.86	L. Evans	USA	1968
800 m		1	42.4	S. Coe	GB	1979
1000 m		2	13.9	R. Wohlhuter	USA	1974
1500 m		3	32.1	S. Coe	GB	1979
1 mile		3	49.0	S. Coe	GB	1979
2000 m		4	51.4	J. Walker	NZ	1976
3000 m		7	32.1	H. Rono	Kenya	1978
5000 m		13	08.4	H. Rono	Kenya	1978
10 000 m		27	22.5	H. Rono	Kenya	1978
20 000 m		57	24.2	J. Hermens	Netherlands	1976
20 944 m	1	00	00.0	J. Hermens	Netherlands	1976
25 000 m	1	14	11.8	W. Rodgers	USA	1979
30 000 m	1	31	30.4	J. Alder	GB	1970

WOMEN

Distances	*Time* *min*	*sec*	*Name*	*Nation*	*Year*
100 m		10.88	M. Oelsner	GDR	1977
200 m		21.71	M. Koch	GDR	1979
400 m		48.60	M. Koch	GDR	1979
800 m	1	54.9	T. Kazankina	USSR	1976
1500 m	3	56.0	T. Kazankina	USSR	1976
1 mile	4	22.1	N. Maracescu	Rumania	1979
3000 m	8	27.1	L. Bragina	USSR	1976

World records for swimming

Distance	Time min	Time sec	Name	Nation	Year
MEN					
100 m		49.44	J. Skinner	USA	1976
200 m	1	49.83	S. Kopliakov	USSR	1979
400 m	3	51.41	V. Salnikov	USSR	1979
800 m	7	56.49	V. Salnikov	USSR	1979
1500 m	15	02.40	B. Goodell	USA	1976
WOMEN					
100 m		55.41	B. Krause	E. Germany	1978
200 m	1	58.43	C. Woodhead	USA	1979
400 m	4	06.28	T. Wickham	Australia	1978
800 m	8	24.62	T. Wickham	Australia	1978
1500 m	16	06.63	T. Wickham	Australia	1979

(2) The alleged weight of the Soviet weightlifter, Vasili Alexeev (world record holder of the super-heavyweight jerk lift), is 162.5 kg. How does this reflect on the validity of the handicapping models proposed?

2 VOTING MODELS

2.1 Problem

In the present British system of voting for Members of Parliament, it is very rare that the % votes cast for each party corresponds to the % of seats gained. Figures for several general elections are given below.

1960	Votes	Seats
Labour	13 064 951	363
Conservative	11 418 433	253
Liberal	2 327 533	12
Scottish Nationalist	128 474	0
Communist	62 112	0
Plaid Cymru	61 071	0
Others	201 032	1
Total	27 263 606	629

1970	Votes	Seats
Conservative and Ulster Unionist	13 144 692	330
Labour	12 179 166	287
Liberal	2 117 638	6
Scottish Nationalist	306 796	1
Plaid Cymru	175 016	0
Communist	38 431	0
Others	383 068	5
Total	28 344 807	629

February 1974	Votes	Seats
Conservative	11 868 906	296
Labour	11 639 243	301
Liberal	6 063 470	14
Scottish Nationalist	632 032	7
Plaid Cymru	171 634	2
Communist	32 741	0
Others	207 884	3
Total	30 615 910	634†

(† including 11 Ulster Unionists)

October 1974	Votes	Seats
Labour	11 456 597	319
Conservative	10 464 675	276
Liberal	5 346 800	13
Scottish Nationalist	839 628	11
Plaid Cymru	166 321	3
Others	195 065	2
Total	28 469 086	634†

(† including 10 Ulster Unionists)

There are several noteworthy points. For example:

(i) despite a dramatic upsurge in Liberal support in February 1974 elections, very few extra seats were won by the party,

(ii) despite polling the greatest number of votes in the February 1974 election, the Conservatives did not win most seats.

Results of this type have been used to argue for a proportional representation system of voting. P. R. aims to ensure that

'% votes for a party = % seats gained in Parliament',

e.g. if the Liberals win 20% of the votes, then they should also win 20% of the seats.

Many countries employ different types of P.R. voting systems, and in order to understand *some* of the problems involved in using a P.R. system, we look at a hypothetical election in which votes cast in a 'super' constituency which returns 5 MPs are as follows.

Party A	199 000
Party B	127 500
Party C	124 000
Party D	49 500

Design an efficient system of P.R. which can be used to allocate (fairly) the 5 seats in the above constituency.

2.2 Solutions

The total votes cast is 500 000, and so each seat is worth 100 000 votes. Using an *exact* form of P.R. we would then allocate

A : 1.99 seats
B : 1.275 seats
C : 1.24 seats
D : 0.495 seats

Unfortunately, seats can be allocated only in whole numbers! Rounding up and down to the nearest integer gives the following.

A : 2 seats
B : 1 seat
C : 1 seat Total *4 seats!*
D : 0 seats

The method clearly does not, in general, work! There are, though, a number of systems which do, and we will describe three of them here.

2.2.1 *Greatest remainder*

Using the quota 100 000 for each seat, the greatest remainder method first allocates a seat for each complete quota obtained, and the rest of the seats are allocated to the parties with the greatest numerical remainder, having subtracted the full quota seats. In this way, we obtain the table below:

Party	Votes Won	Whole Seats Won	Remainder	Extra Seats	Total
A	199 000	1	99 000	1	2
B	127 500	1	27 500	0	1
C	124 000	1	24 000	0	1
D	49 500	0	49 500	1	1

2.2.2 *d'Hondt rule*
This method was devised by a Belgium academic, Victor d'Hondt, and is extensively used in Europe. Dividing the votes total by 1, 2, 3 . . . gives the table below:

Divided by	A	B	C	D
1	199 000	127 500	124 000	49 500
2	99 500	63 750	62 000	24 750
3	66 333	42 500	41 333	16 500
4	49 750	31 875	–	–

The 5 seats are now allocated to the 5 highest numbers on the table (these are ringed).

A : 3 seats
B : 1 seat
C : 1 seat
D : 0 seats

2.2.3 *Scandinavian compromise*
This is essentially similar to the d'Hondt rule, except that the divisors are 1.4, 3, 5, 7, . . . instead of 1, 2, 3, 4, ... Using this method the 5 highest figures give the following result.

A : 2 seats
B : 2 seats
C : 1 seat
D : 0 seat

2.3 Remarks

The summary of the three systems used is given below.

Seats Won	Greatest remainder	d'Hondt rule	Scandinavian Compromise
A	2	3	2
B	1	1	2
C	1	1	1
D	1	0	0

Rather surprisingly, the three methods all give different results! It has been found that the 'greatest remainder' method favours the smaller parties (for example, party D above), the 'd'Hondt rule' helps the strongest party (for example, party A above), whilst the 'Scandinavian compromise' gives little help to smaller parties, but also does not let the strongest party dominate (for example, party A above does not get a third seat, and party D gets no seat.

For the problem (as we have set it) there is clearly no perfect solution, and many countries have adopted far more complicated system of P.R. For example, the West Germans have developed a unique electoral system called 'the double vote', which is based both on the British 'single-member' constituencies and on our overall method of proportional representation. Of the 496 members of the Bundestag (the lower house) exactly half are directly elected constituency members. A West German voter has two votes, namely:

(i) one for a list of names and parties (just as in the British system),
(ii) one for the party of his choice.

The second vote decides the overall proportion of seats that each party will finally secure, and so a voter has every opportunity to vote for a personality rather than party for their first vote, and this will not affect the final balance of the parties. For example, in the 1972 Federal election the 248 *constituency* seats resulted in

Social Democrats	152
Christian Democrats	96
Free Democrats	0

whereas the second vote (calculated over the whole of West Germany) resulted in the following percentage vote

Social Democrats	45.8%
Christian Democrats	44.9%
Free Democrats	8.4%

This means that the Social Democrats, for example, should have an overall total of about 230 seats in the Bundestag. Having obtained 152 constituency seats, the remaining 78 are taken from the parties 'list' of candidates. The actual results were:

	Seats			
	Constituency	Party List	Total	%
Social Democrats	152	78	230	46.3%
Christian Democrats	96	129	225	45.4%
Free Democrats	0	41	41	8.3%

Although this system does have some obvious advantages (particularly for the smaller parties such as the Free Democrats), it does have one snag. It is quite feasible for a party to win *too* many constituency seats (for example, if the Social Democrats had won more than 230 constituency seats in the 1972 election). If this happens, then there has to be a temporary increase in the number of seats in the Bundestag. In 1961, for example, there were five extra members. As well as this mathematical disadvantage, it has the political disadvantage of making the parties all-important, because of the party list system of allocating half of the members.

2.4 Related Problems

(1) Apply the three methods of P.R. described above for your neighbouring city, by combining all the n individual city constituencies together to form a super constituency which returns n MPs, and use the total votes obtained for each party in all the n constituencies at the last general election (voting figures are given in *Whitaker's Almanack*).

(2) In a two-party election using the 'first past the post' method of electing MPs, it has been postulated that

$$y = \frac{x^3}{3x^2 - 3x + 1},$$

where y is the proportion of seats won by one of the parties and x the corresponding proportion of votes. Does this seem a reasonable model? Find suitable data and test the model.

3 ATTRACTIVENESS OF COMPETING SHOPPING CENTRES

3.1 Problem

One of the difficulties faced by planners of new shopping centres is to estimate the likely success of the venture. Will it produce sufficient customers for the retail shops to make a profit, how large should the centre be, how much car parking space should be provided, from how far away will it attract shoppers, etc?

In many new large shopping centres, the parking facilities have been found to be far from adequate in peak shopping times. One of the reasons given for this is that shoppers are willing to travel surprisingly long distances to shop at attractive new centres, which offer a full and comprehensive range of shops. The problem that we pose here is how to estimate the regions of influence between *competing* shopping centres.

For example, suppose Northamptonshire County Council is proposing to build a new shopping centre at a site midway between the towns of Northampton and Wellingborough. Northampton already has a range of attractive shops, but parking is very difficult and there is little chance for any major expansion. On the other hand, Wellingborough lacks many of the popular shops, but these retailers would be unwilling to move there as they would feel that the population of Wellingborough would not be sufficient to make a reasonable profit.

Estimate the zone of influence of the proposed new shopping centre; that is, determine the area from where it is more probable that inhabitants will shop at the new centre, rather than the established centres at Northampton and Wellingborough.

3.2 Possible Solution

There are many ways in which to tackle this problem, and the method we propose below must not be regarded as *the* answer – it is just one possible mathematical approach to the problem. You might well achieve a better solution without mathematics!

We firstly consider just the zone of influence of the two established centres at Northampton and Wellingborough. Let T be the probability of an individual making a trip to a particular shopping centre. What will T depend on? We will assume that T depends on two key parameters,

(i) D, the distance of the individual from the centre,
(ii) F, the measure of the intrinsic attractiveness of the centre.

The parameter D depends on where the individual lives, whilst the parameter F depends on the amenities of the centre. So D is quite easily measurable, whilst F is dependent on many factors, for example

(i) the number of shops in the centre and their retail floor space,
(ii) the number and type of departmental stores (e.g. Marks & Spencer, John Lewis),
(iii) the number of supermarkets,
(iv) other attractions in the centre (e.g. libraries, museums, cinemas, parks).

It is also clear that different individuals will probably give different ratings for F!

Now Northampton, serving a local population of over 100 000, has a large shopping centre which includes a Debenhams department store and a branch of

bypass the centre at B and shop at the centre A, which is further away in distance but with more attractive amenities.

Let us apply this theoretical model to our practical example. We earlier introduced the relative attractiveness between the towns of Northampton and Wellingborough for which we estimated $c = F_1/F_2 = 3$. This is the case illustrated in Fig. 3.2, and superimposing this diagram on a map of the area around Northampton and Wellingborough results in Fig. 3.3.

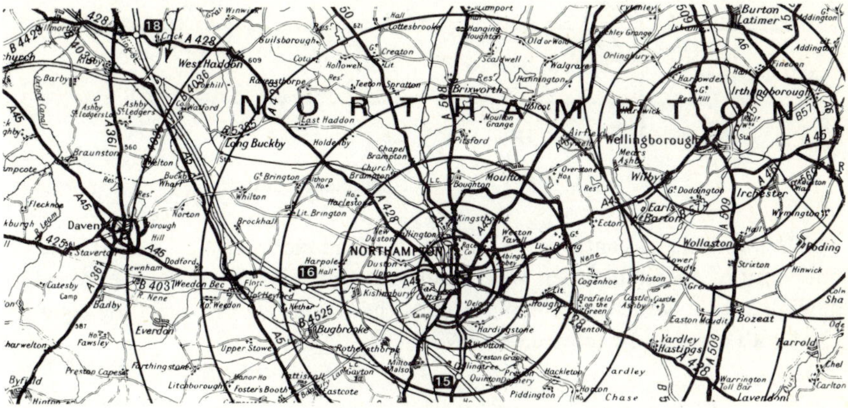

Fig. 3.3 – Regions of Influence of Shopping Centres.

According to the model a number of villages such as Bozeat and Sywell, which are in fact closer in distance to Wellingborough, are principally attracted to Northampton. It would be interesting to try to validate the model by conducting surveys in villages which are near the boundary of attractiveness between the two centres. For inhabitants without cars, it will probably be found that the most significant factor is not the attractiveness of centres or distances, but the availability and frequency of bus services into the shopping centres.

Having developed a method of determining zones of influence, we can now return to our original problem of introducing the new centre at a site midway between Northampton and Wellingborough. The zone of influence will obviously depend on its attractiveness relative to the two established sites. For example, if we rate the proposed centre as twice as attractive as Northampton (and so six times more attractive than Wellingborough), it is easy to determine the boundaries of choice between the three shopping centres. If the proposed site for the new centre is situated midway between the two towns, we have the situation in Fig. 3.4. Points P on the boundary of choice between Northampton and the new centre satisfy equation (3.4) with $c = 1/2$ and a replaced by $a/2$, i.e.

$$\left(x + \frac{a}{2}\right)^2 + y^2 = \frac{a^2}{2} \quad ,$$

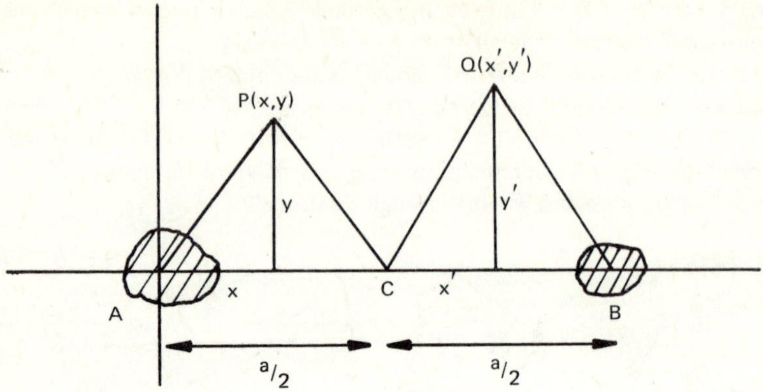

Fig. 3.4 – Effect of Hypermarket at C.

which is a circle of centre $x = -a/2$, $y = 0$ and radius $a/\sqrt{2}$. To determine the boundary of choice between Wellingborough and the new centre, we introduce new axes x', y' with C the new centre as the origin. We can again use equation (3.4) with $c = 6$ and a replaced by $a/2$, i.e.

$$\left(x' - \frac{3}{5}a\right)^2 + y' = \frac{3a^2}{50} \quad ,$$

which is a circle centre $x' = \frac{3}{5}a$ (which means $x = \frac{11}{10}a$) and radius $\sqrt{\frac{3}{50}}\,a$.

The two boundaries are illustrated in Fig. 3.5 on the Northampton/Wellingborough map. While not providing actual predictions about the number of shoppers, it does clearly show the devastating effect that the proposed new centre would have on the shops in Wellingborough. It predicts that people in

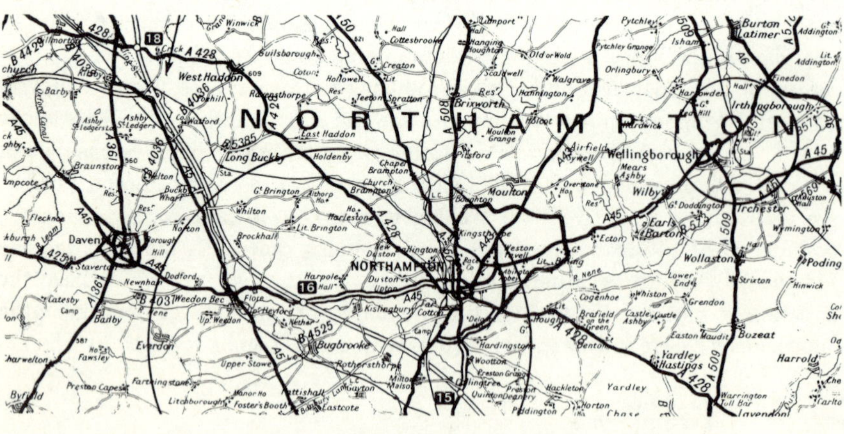

Fig. 3.5

such places as Wollaston, Rushden, and even Irthingborough would have preference for the proposed new centre.

It also predicts that inhabitants who live on the east side of Northampton will be attracted to the centre. This is vitally important for the success of the centre. Prestigious retailers will agree to take sites in the new centre only if they are convinced that the new centre will be well patronised. This is the basic dilemma of the planner – can he convince the important retailers (John Lewis, Marks & Spencer, Boots, British Home Stores, Woolworths, *et al.*) that the centre will really attract the needed customers? Although mathematical models of this type can help in understanding the situation, it is not so easy to be convinced that they can be used for accurate predictions.

3.3 Related problems

(1) Apply this model to two neighbouring shopping centres in your vicinity. Determine a suitable value for c, the ratio of attractiveness of the centres, by taking the average value of your friends' estimates for this value.

(2) Analyse the problem of three centres, situated at the vertices of an equilateral triangle. Firstly consider the problem if the three centres are all rated equally attractive, then generalise by taking, for example, one of the centres to be twice as attractive as the other two.

(3) Rework the problem using the following forms for the functional form of T.

(i) $T = KFe^{-\lambda D}$

(ii) $T = KF/D$.

4 BREAK-EVEN ANALYSIS

4.1 Problem

A motor vehicle manufacturer is running short of a particular vehicle component because of a strike at the normal supplier. The company approaches a number of small engineering companies and asks if they can supply batches of the component to make up the deficit.

Suppose that you are the sales manager of one of these companies. The vehicle manufacturer supplies the component specification and is prepared to pay £2 per component delivered by a certain date, and asks if you are prepared to accept the deal, and if so how many units you can supply.

You know that it will cost £5000 to set up the specialised equipment for the job, and that you could produce (by the stipulated date) a maximum of 13 000 units at a cost of £1.60 per unit in addition to the set-up cost. Would you accept the offer, and if so how many units would you agree to supply?

Having answered the question, you will probably realise that there could be many situations where the same kind of analysis should be done. If you have not already done so, set up a formal mathematical model governing situations of this type and apply it to the specific case above.

4.2 Possible solution

REPORT

Prepared by: Mr. T. Percent A.A.A. Consultants Ltd.

Client: Vehicle Products Ltd.

Subject: Oil Filters for L.B. Cars Ltd.

Preamble

The advice outlined below is based on information and data given by you at our meeting of 16th October 1978. Our subsequent analysis of the situation forms an appendix to this brief report.

Conclusions

General:

To break even on a venture of this type Vehicle Products Ltd. (V.P.) must be guaranteed sufficient volume of production to offset set-up costs. Our analysis (see appendix) indicates that the break-even point is reached when

$$\text{b.e. volume of production} = (\text{set-up costs})/\text{marginal return}$$

and thereafter

$$\text{profit} = (\text{marginal return}) \times (\text{volume of production}) - \text{set-up costs.}$$

Specific to this opportunity:

Break-even volume is 12 500 units. Even at a guaranteed volume of 13 000 units, profit would be £200.

This profit represents a return of 4% on V.P.'s investment of £5 000.

Recommendation:

Whether this represents a sufficient return on capital invested *must be a matter of your commercial* judgement but we *would not recommend an investment giving less than 18% return* unless the timescale involved is exceedingly short.

APPENDIX: MATHEMATICAL MODEL

Stage 1 – The Real World

When a company produces a product it incurs costs. The company sells the product at a fixed price and is interested in examining its profitability.

Stage 2 – Formulation

We suppose the costs are of two types: fixed and variable. The fixed costs are independent of the number of units produced (e.g. rent and rates), while the variable costs increase with the number produced (e.g. materials). Initially we assume that the variable costs are directly proportional to the number of units produced – this should simplify our model. The company has a certain amount of money coming in from sales, and wants to ensure that profit is maximum. For convenience, we assume that all units produced are sold immediately.

Stage 3 – The Model

Let x be number of units produced and sold,
C be total cost of production (£),
I be income from sales (£),
P be the profit (£).

Our assumptions above state that C consists of two parts:

fixed cost a (£),
variable cost b (£/unit produced).

Then $$C = a + bx \quad . \qquad (1)$$

Also, income I depends on selling price s(£/unit).

Thus $$I = sx \qquad (2)$$

The profit P is then the difference between income and costs,

i.e. $$\begin{aligned} P &= I - C \\ &= sx - (a + bx) \\ &= (s - b)x - a \quad . \qquad (3) \end{aligned}$$

We now have a mathematical model consisting of the RELATIONSHIPS (1) – (3) between the VARIABLES x, C, I, P, a, b, s. These variables may be classified as:

independent (or decision)	x
dependent	C, I, P
parameters	a, b, s .

Stage 4 – Analysis

The manufacturer, knowing x, a, b, s, can determine P. He can also see that to break even (i.e. make neither profit nor loss) he must produce $x = a/s - b$ units.

The model is best summarised graphically.

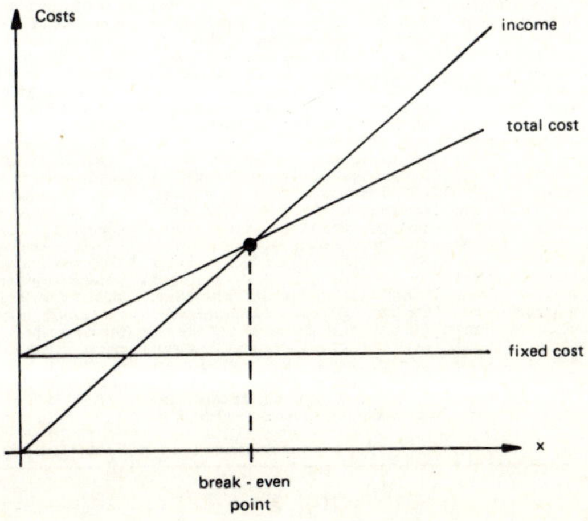

Stage 5 – Comparison with Reality

The model agrees with out intuition, in that if few units are sold a loss will result but if lots of units are sold a profit will result. If the break-even point proves to be unrealistic, then a nonlinear model could be tried or our simplifying assumptions about cashflow amended.

4.3 Comment

A cutting from the newspaper, showing an actual example of this analysis, is given in Fig. 4.1.

Shell profits from paradox

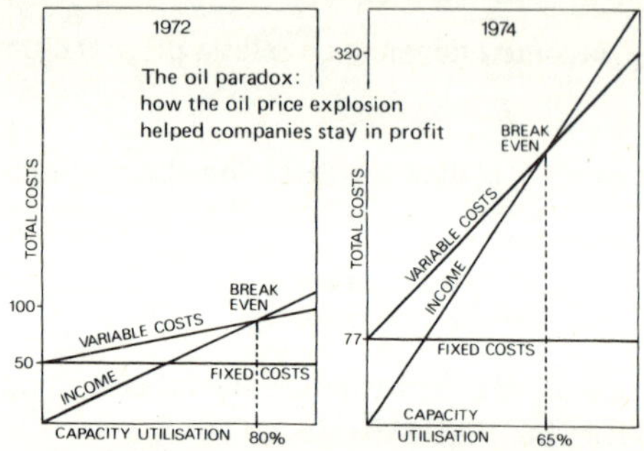

Comparative cost models for a 450,000 ton/year olefins plant in 1972 and 1974

IT WAS SHELL'S quarterly figures in the spring of this year that set city analysts off in search of the X factor. Just what was behind a profits performance better than all but the most optimistic had expected? How could Shell and BP manage to avoid severe losses in some of their operations when demand and throughput were at a lower level than for many a year?

Now, the answer is beginning to emerge. And the explanation is likely to spark off a reappraisal of the shares in our two biggest oil firms, as well as ICI and other chemicals groups.

Shell chairman Sir Frank McFadzean held out the key to a fresh perspective on profitability in his annual statement to shareholders.

"While the effect of inflation on the group financial results for 1974 was substantial, even more important was the impact of massive increases in crude oil prices," he said. Among the first to spot the deep significance of the remark were the investment research team at stockbrokers Hoare & Co Govett. Others were soon reworking their own profit projections afresh, and the end result will be a good deal more big investor interest in oils and chemicals.

What the stockbrokers spotted was that massive inflation of raw materials and other variable costs changes the economics of many chemical plant operations. The net result, as we show below, is to reduce breakeven throughput substantially so that profits can be made at levels of production which, a few years back, would almost certainly have resulted in losses.

Conventional wisdom has been that oil refineries, because of their huge capital investment, could barely live with throughput as low as 65% of maximum capacity. But some figures compiled by Esso show that because inflation increased the costs of raw materials and other variables relative to fixed costs, some basic parts of their business could break even at 65% of capacity, whereas a couple of years earlier at least 80% was needed.

It was time someone found some less than totally disastrous side effects of inflation just to break the monotony. Anyway, here is what happened to the cost structure of certain of Esso's plants used for the basic petrochemical process of steam-cracking naphtha to produce olefins.

In 1972 the basic variable costs, feedstock, fuel and utilities amounted to 47% of the total bill for running the plant. The fixed costs, mainly labour, depreciation and a factor representing the return on capital tied up in the plant, made up the rest. Just two years later, all of that had changed beyond recognition. The "fixed" costs had pushed up by 45%, while the variables rocketed by 5.3 times due to the increased price of crude oil. As our chart shows, the fixed costs become a much smaller proportion of the total and are recovered more easily (assuming normal pricing policies). The breakeven point of the operation falls.

This goes some way to explaining the widespread surprise which greeted Shell's first quarter income revealed in the middle of May. This change in the basic mix of costs has some other important implications for major British companies, especially those using oil heavily.

- Economies of scale are not so important as they once were. Since break-even points are lower, sheer size might be counterbalanced by increased transport costs or other factors.
- Access to cheap raw materials becomes more important relatively than access to cheap plant. In the past BP had cheaper crude oil but higher refining costs than Shell. That situation may now be reversed.
- Because of a near 400% rise in the price of phosphate rock since late 1973, the economics of Albright & Wilson's troublesome plant in Newfoundland can now make sense even if it runs at far below maximum capacity.
- Similar effect will be felt in the papermaking industry and other businesses with a high proportion of fixed to variable costs, and using raw materials whose prices have multiplied.

All of this makes shares like BP Shell and ICI much more likely to wheather the storm of the gathering industrial recession and come through the trough relatively unscathed. I think perhaps BP may be the biggest gainer here in so much as the trough promised to have a particularly severe impact.

JOHN BELL

Reproduced from the *Sunday Times Business News* 13/7/75

Fig. 4.1

5 SHORTEST STOPPING DISTANCES

5.1 Problem statement

The following information on stopping distances is reproduced from the Ministry of Transport *Highway Code*.

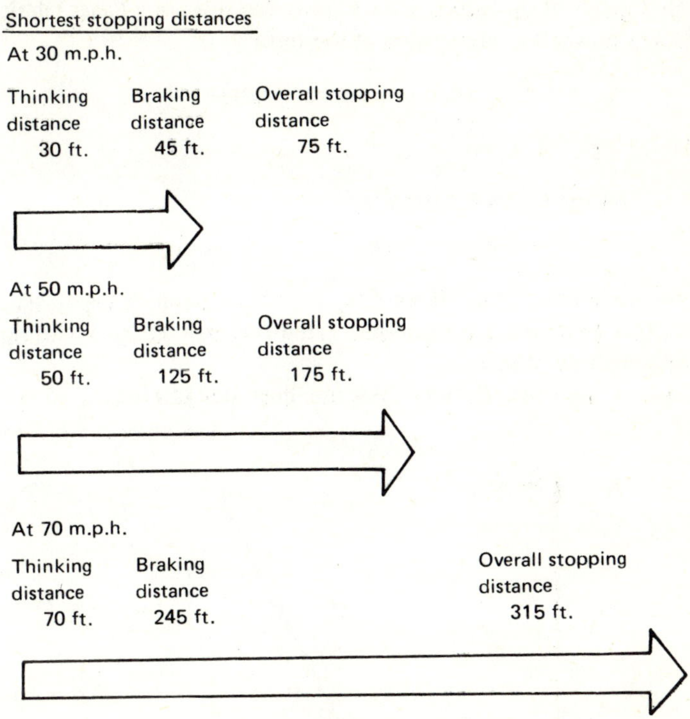

Fig. 5.1

Can you:

(i) find an empirical model (i.e. formula) which fits the data,

(ii) produce a theoretical model which gives an explanation for the formula obtained in (i) ?

5.2 Empirical model

The data from the *Highway Code* divide the shortest stopping distance, say d, into two components, namely

(i) the THINKING DISTANCE d_T (which is the distance travelled whilst the driver's foot moves from the accelerator to the brake pedal),

(ii) the BRAKING DISTANCE d_B (which is the distance travelled whilst braking from speed v to 0).

Looking at the data from the *Highway Code*, it is clear that the thinking distance is modelled by

$$d_T = v \text{ feet} \quad . \tag{5.1}$$

It is not quite so clear what formula has been used for the braking distance, although a graph of d_B against v does show that it is not a linear relationship. If we assume a power law relationship of the form

$$d_B = k\,v^{\alpha} \qquad (k, \alpha \text{ positive constants})$$

then, taking logs, we obtain

$$\begin{aligned} \log d_B &= \log k + \log v^{\alpha} \\ &= \alpha \log v + \log k \end{aligned} \tag{5.1'}$$

using the properties of logs. If we now consider a graph of $\log d_B$ against $\log v$, equation (5.1′) predicts a straight-line graph (i.e. $y = \alpha x + c$) with slope α and intercept c with the y axis.

The data from the *Highway Code* are illustrated in a $\log d_B - \log v$ graph in Fig. 5.2.

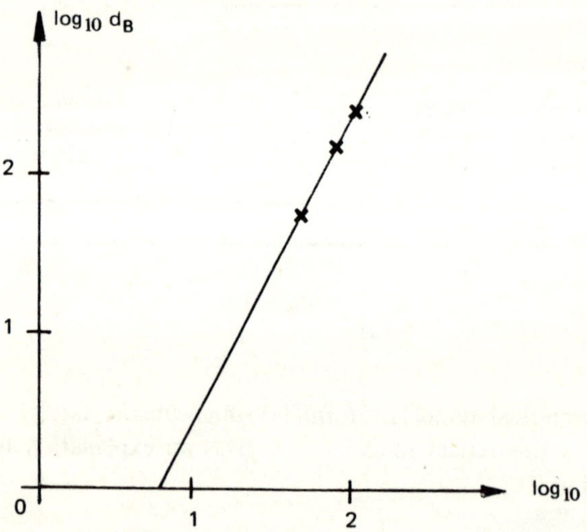

Fig. 5.2

We do indeed obtain a straight line, with slope 2. Also, from the intercept with the $\log d_B$ axis, we obtain $k = 0.05$. Hence a possible model used for this data is

$$d_B = v^2/20 \text{ feet.} \tag{5.2}$$

So, combining (1) and (2), we clearly have the formula

$$d = v + v^2/20 \text{ feet} \quad , \tag{5.3}$$

where v is the speed in m.p.h.

5.3 Theoretical Model

We again divide the stopping distance into two parts.

(i) *Thinking distance,* d_T

The thinking distance is the distance travelled whilst the driver is deciding to apply the brake pedal and actually moving his/her foot over to the brake pedal. If the time taken to do this is T, then, since the speed will remain approximately constant, we have $d_T = v\,T$.

Now if v is in m.p.h. then

$$v \text{ m.p.h.} = v \,.\, \frac{22}{15} \text{ feet per second,}$$

giving

$$d_T = \frac{22}{15} v\,T \quad .$$

To produce the empirical formula (5.1), we must take $T = 15/22$ seconds! Is this reasonable? Racing drivers probably have a much shorter reaction time, whilst others might have a value well in excess of 15/22 seconds.

The important point to note is that the thinking distance travelled increases *linearly* with the speed – double the speed, and the thinking distance doubles. We do not have any great confidence that the actual values quoted are particularly significant – the linear dependence is the vital message for drivers.

(ii) *Braking distance,* d_B

It is much more difficult to know how to start here. For most of this part of the motion full braking power is being supplied, and the simplest assumption is to postulate that the braking force applied to the car is a constant multiple of the car's mass. Then Newton's second law of motion gives

$$m \frac{d^2x}{dt^2} = -k\,m\,g \quad , \qquad k \text{ constant,}$$

where g is acceleration due to gravity and x is distance travelled whilst braking. Integrating,

$$dx/dt = -k\,g\,t + A$$

and at $t = 0$, say, $dx/dt = v$, giving

$$dx/dt = -k\,g\,t + v \quad . \tag{5.4}$$

Integrating again

$$x = -\frac{1}{2} k g t^2 + v t + B$$

and, since $x = 0$ at $t = 0$, we have $B = 0$. Thus

$$x = -\frac{1}{2} k g t^2 + v t \ . \tag{5.5}$$

This gives the distance travelled whilst braking. We require the total braking distance, which is given by (5.5) when the time t is given by (5.4) with $dx/dt = 0$, i.e.

$$t = v/kg$$

and so

$$d_B = -\frac{1}{2} k g \left(\frac{v}{kg}\right)^2 + v \left(\frac{v}{kg}\right) ,$$

i.e

$$d_B = \frac{1}{2} \frac{v^2}{kg} \ . \tag{5.6}$$

Now if we want d_B in feet, and v is in m.p.h. then we must substitute v in feet per second into (5.6) and $g = 32.2$ ft/s^2.
Thus

$$d_B = \frac{\left(\frac{22\,v}{15}\right)^2}{64.4\,k}$$

We are now in a 'chicken and egg' situation. To obtain the empirical formula (5.2), we need the appropriate value of k; but the actual value of k depends on the construction of the car! In fact, $k = 2/3$ leads to

$$d_B = v^2/19.96 \ ,$$

which is approximately the same as (5.2). Is $k = 2/3$ reasonable? This is not clear, but the Ministry of Transport is happy to use this value, and, as in (5.1), the important point to make is the v *squared* dependence of the braking distance on the speed and not the precise values, so a doubling of the speed gives a fourfold increase in the braking distance.

5.4 Related Problems

(1) Evaluate the braking distance with $k = 0.5, k = 1.0$. Are the braking distances significantly altered?

(2) Can you suggest a better way of displaying the shortest stopping distances data in the *Highway Code?*

6 MAXIMISING TRAFFIC FLOW THROUGH TUNNELS

6.1 Problem

Many road tunnels in Great Britain suffer severe road congestion during busy work periods (normally 8-9 a.m. and 5-6 p.m.). To maximise the flow of traffic during a busy period, many tunnel authorities prescribe recommended speeds and separation distances. What recommendations would you make?
[*Note* You might find the data from the case study *Shortest stopping distances* of some use.]

6.2 Solution

The important concept to start with is the FLOW RATE, which we define as the number of cars passing a fixed point in a unit time interval. The flow rate will depend on a number of factors, such as

(i) traffic speed,
(ii) separation distance between cars,
(iii) length of cars.

For simplicity, we assume a uniform stream of traffic, moving with speed v m.p.h., with an average separation distance d feet between vehicles and an average car length b feet. This is illustrated in Fig. 6.1.

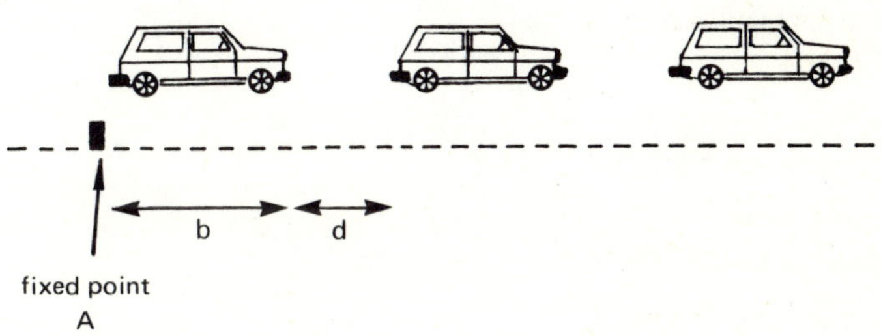

Fig. 6.1 – Uniform stream of traffic.

We first note that v m.p.h. $= \frac{22}{15} v$ feet per second. Now consider a fixed post, A, and suppose the back of a car has just passed this post (see Fig. 6.1). We want to find the time taken for the next car in the stream to pass A; this is given by

$$\frac{\text{distance yet to travel}}{\text{speed}} = \frac{d+b}{22\,v/15}\ .$$

Now the flow rate is the number of cars passing A in unit time,

i.e. $$\text{flow rate} = \frac{1}{\text{time taken by each car}}$$

$$= 1 \Bigg/ \left[\frac{d+b}{22\,v/15}\right] ,$$

i.e. $$f = \frac{22\,v/15}{b+d} . \tag{6.1}$$

Thus the flow rate depends on both speed v and separation distance d, so the traffic manager of H.T. A. has *two* possible controls, namely, speed v and separa-separation distance d.

The speed v, however, clearly affects the distance d. The data given on the traffic manager of the tunnel has *two* possible controls, namely, speed v and shortest stopping distance, d is made up of two parts, the thinking distance

$$d_{\mathrm{T}} = v \text{ feet} \tag{6.2}$$

and the braking distance

$$d_{\mathrm{B}} = v^2/20 \text{ feet} . \tag{6.3}$$

So our problem is now to decide what formula to take for the separation distance, d. We start looking at two extreme cases, by taking d as just the thinking distance (many cars on motorways drive in this mode), and secondly taking d as the complete stopping distance $d_{\mathrm{T}} + d_{\mathrm{B}}$.

(i) $d = d_{\mathrm{T}}$

In this case

$$f = \frac{22\,v}{15\,(v+b)} , \tag{6.4}$$

and this is illustrated in Fig. 6.2.

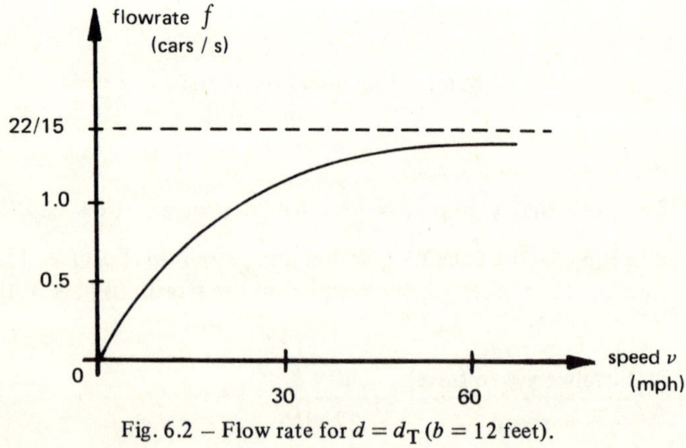

Fig. 6.2 – Flow rate for $d = d_{\mathrm{T}}$ ($b = 12$ feet).

We see that $f \to 22/15$ as $v \to \infty$, and the policy seems to be 'the faster the better'! This is certainly not the policy to be encouraged by any tunnel authority. It is, though, interesting to note that the flow rate increases only very slowly above 30 m.p.h.

(ii) $d = d_T + d_B$

Here the flow rate is given by

$$f = \frac{22\,v}{15\,(v + v^2/20 + b)} \ . \tag{6.5}$$

The behaviour of this function is illustrated in Fig. 6.3.

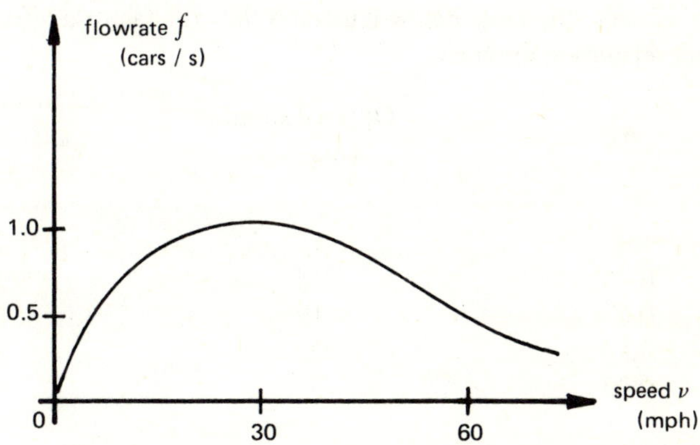

Fig. 6.3 – Flow rate for $d = d_T + d_B$ (b = 12 feet).

This function has a maximum value at a finite value of v. It is given by the solution of

$$df/dv = 0 \ ,$$

i.e.
$$(v + v^2/20 + b) = v\,(1 + v/10) \ . \tag{6.6}$$

This gives $v = \sqrt{(20\,b)}$ and with b = 12 feet, we obtain $v \approx 16$ m.p.h. The corresponding separation distance is 28 feet. Again, this extreme case is probably not a very realistic one. Most drivers will want to drive faster than 16 m.p.h., and closer than 28 feet!

To look for a situation somewhere between these two extremes, we take the separation distance as

$$d = d_T + \alpha\,.\,d_B \ , \tag{6.7}$$

where α is a parameter in the range $0 < \alpha < 1$. In this case

$$f = \frac{22\,v}{15\,(v + \alpha\,v^2/20 + b)} \quad . \tag{6.8}$$

Now $df/dv = 0$ gives

$$1.\,(v + \alpha\,v^2/20 + b) = v\,(1 + \alpha\,v/10) \quad ,$$

i.e. $$v = (20\,b/\alpha)^{1/2} \quad .$$

The corresponding separation distance is

$$v + \alpha v^2/20 = v + b \quad .$$

For $b = 12$ feet, the table below illustrates how the factor α affects the optimal speed and separation distance.

α	Optimal speed (m.p.h.)	Corresponding separation distance (feet)
1.0	15	27
0.75	18	30
0.67	19	31
0.5	22	34
0.25	31	43

A possible solution for the problem is to display a notice board at the entrance of the tunnel stating:

IN CONGESTED TRAFFIC
TRAVEL AT 20 m.p.h.
SEPARATION DISTANCE 11 YARDS

One important point to note is that as well as specifying the optimal speed, the recommended separation distance must also be given, otherwise operating conditions may be far from optimal!

6.3 Related problems

(1) How sensitive are the results obtained in section 6.2 to
  (i) the value used for b, the average car length,
  (ii) the value used for the thinking time?

(2) What suggestions would you make (if any) for regulating the flow of congested traffic on a multilane motorway?

7 RETAIL PRICE INDEX

7.1 Problem statement

In a period of high inflation the prices of most goods tend to rise, but the actual *rate* at which prices increase varies a good deal. (During the severe German inflation of 1920–23, prices increased by more than a thousand million times.)

In Britain, the Government publishes each month the *Retail Price Index* which measures the general level of prices. It is calculated by choosing a list of goods bought by the 'average' family and which cost £100 at a particular date. The same (or equivalent) goods are then costed each month. The list is supposed to be typical for an average family and would include food, housing, heating, travel, rates, etc.

The official table below shows how the price index has changed over a number of years, taking the January 1974 cost as £100; if Mr and Mrs Average spent £100 on their shopping in January 1974, then to buy the same goods in January 1964 would have cost them £54.60, and to buy them in January 1976 would have cost £147.90.

	Jan.	Feb.	Mar.	Apr.	May	June	July	Aug.	Sep.	Oct.	Nov.	Dec.
1963	53.5	54.0	54.1	54.2	54.2	54.2	53.9	53.7	53.9	54.1	54.2	54.3
1964	54.6	54.6	54.8	55.3	55.8	56.0	56.0	56.2	56.2	56.3	56.7	56.9
1965	57.1	57.1	57.3	58.4	58.6	58.8	58.8	58.9	58.9	59.0	59.2	59.5
1966	59.6	59.7	59.7	60.5	60.9	61.1	60.8	61.2	61.1	61.2	61.6	61.7
1967	61.8	61.8	61.8	62.3	62.3	62.5	62.2	62.0	61.9	62.4	62.8	63.2
1968	63.4	63.7	63.9	65.1	65.1	65.4	65.4	65.5	65.6	65.9	66.1	66.9
1969	67.3	67.7	67.9	68.7	68.6	68.9	68.9	68.7	68.9	69.5	69.6	70.1
1970	70.6	71.0	71.4	72.5	72.7	72.9	73.5	73.4	73.8	74.6	75.1	75.6
1971	76.6	77.1	77.7	79.4	79.9	80.5	80.9	81.0	81.1	81.5	82.0	82.4
1972	82.9	83.3	83.6	84.4	84.8	85.4	85.6	86.3	86.8	88.0	88.3	88.7
1973	89.3	89.9	90.4	92.1	92.8	93.3	93.7	94.0	94.8	96.7	97.4	98.1
1974	100.0	101.7	102.6	106.1	107.6	108.7	109.7	109.8	111.0	113.2	115.2	116.9
1975	119.9	121.9	124.3	129.1	134.5	137.1	138.5	139.3	140.5	142.5	144.2	146.0
1976	147.9	149.8	150.6	153.5								

This table, however, is about the past; the problem is to find a method of estimating what price changes are likely in the future. It is of the utmost importance for government planning to be able to forecast the likely level of price increases in the coming months. It is also vital to both management and union leaders in their negotiations on wage settlements.

Using the data in the table above, devise a method of predicting the Retail Prices Index and test your method by predicting the value of the Retail Price Index for each of May 1976, June 1976, July 1976.

7.2 Possible solutions

Model 1

The simplest model used for forecasting is to assume that in the short term the underlying value is constant, but subject to random variations. The constant underlying value may be estimated by averaging successive values of the index. So, for example, we might choose to average the previous six monthly values of the index in order to give an estimate of the underlying value and then use this as our forecast. Applying the model to the data above in order to estimate the July 1963 figure, we obtain the following calculation.

$$\frac{53.5 + 54.0 + 54.1 + 54.2 + 54.2 + 54.2}{6} = \frac{324.2}{6} = 54.0$$

The 6-monthly average gives a forecast of 54.0 which is in good agreement with the actual value of 53.9.

Continuing in this way, always using the previous six monthly values for the average (called the 6-monthly moving average) gives the fourth column in the following table, and these form the forecast. The actual values are given in the final column.

Month	Index	Total in last 6 months	6-month moving average	Actual value
Jan. 1963	53.5			
Feb.	54.0			
Mar.	54.1			
Apl.	54.2			
May	54.2			
June	54.2	324.2	54.0	53.9
July	53.9	324.6	54.1	53.7
Aug.	53.7	324.3	54.05	53.9
Sep.	53.9	324.1	54.0	54.1
Oct.	54.1	324.0	54.0	54.2
Nov.	54.2	324.0	54.0	54.3
Dec. 1963	54.3	324.1	54.0	54.6
Jan. 1964	54.6	324.8	54.1	54.6

Comparing the figures in the last two columns shows that if there is any trend then the forecast will always lag well behind. From July 1963 onwards, the index is slowly rising whereas the *forecast* does not start to increase until December 1963. This fact is even more apparent if we use the model at April 1976 in order to predict the May 1976 value. The 6-monthly moving average at April 1976 is

$$\frac{144.2 + 146.0 + 147.9 + 149.8 + 150.6 + 153.5}{6} = 148.7 ,$$

so the forecast for May 1976 is 148.7. This is hardly likely to be realised, since the April value was already 153.5. Clearly the model must be amended to take account of the trend.

Models 2 and 3

If we were to use, for example, a 5-month moving average on the data, then this is an estimate of the index at the mid-point of the third month. We could continue to calculate the 5-month moving averages, and then estimate the trend by evaluating the differences between successive moving averages (we call this model 2). If the rate of increase is constant, then we would expect these differences to be constant. If we assume, though, that the trend itself is subject to random variation, it is probably advisable to take a moving average of the trend (this yields model 3).

	Month	Index	5-month M.A.	Trend	3-month M.A. of the trend
1975	May	134.5			
	June	137.1			
	July	138.5	138.0	+ 1.6	
	Aug.	139.5	139.6	+ 1.4	1.50
	Sept.	140.5	141.0	+ 1.5	1.53
	Oct.	142.5	142.5	+ 1.7	1.70
	Nov.	144.2	144.2	+ 1.9	1.73
	Dec.	146.0	146.1	+ 1.6	1.80
1976	Jan.	147.9	147.7	+ 1.9	
	Feb.	149.8	149.6		
	Mar.	150.6			
	Apl.	153.5			

To produce the estimate for May 1976, the *moving average* of the index must be projected forward. The latest figure is 149.6, which is the underlying figure at

February 1976. This must be projected forward 3 months to give an estimate for May 1976. Model 2 uses the latest trend value, whilst Model 3 substitutes the latest value of the moving average of the trend.

Thus Model 2 will forecast $149.6 + (3 \times 1.9) \quad = 155.3$
and Model 3 will forecast $149.6 + (3 \times 1.80) = 155.0$

So, continuing with the same value for the trend, we obtain the estimates in the table below, which also includes the actual figures.

Index			
	Predictions		
Month	Model 2	Model 3	Actual
May 1976	155.3	155.0	155.2
June 1976	157.2	156.8	156.0
July 1976	159.1	158.6	156.3

Agreement is reasonable for both models for May 1976, but by July the agreement does not look too good. Continuing in this way, the predictions (at April 1976) for the rest of the year are given below.

Index			
	Predictions		
Month	Model 2	Model 3	Actual
Aug. 1976	161.0	160.4	158.5
Sept. 1976	162.9	162.2	160.6
Oct. 1976	164.8	164.0	163.5
Nov. 1976	166.7	165.8	165.8
Dec. 1976	168.6	167.6	168.0

Rather surprisingly, both models seem to be each in close agreement by the end of the year, and there appears little to choose between the two models.

Models 4 and 5
One obvious feature of the data is the greater monthly increase in the index from March to April in most years (due to budget changes). Since these occur at 12-monthly intervals, it may be better to work with a 12-month Moving Average of the index. (This would also have the advantage of smoothing out any other regular annual variation.)

Month		Index	12-month M.A. to date	Trend	Prediction Model 4	5-month M.A. of trend to date	Prediction Model 5
1974	Apl.	106.1					
	May	107.6					
	June	108.7					
	July	109.7					
	Aug.	109.8					
	Sept.	111.0					
	Oct.	113.2					
	Nov.	115.2					
	Dec.	116.9					
1975	Jan.	119.9					
	Feb.	121.9					
	Mar.	124.3	113.7				
	Apl.	129.1	115.6	1.9			
	May	134.5	117.9	2.1			
	June	137.1	120.2	2.3			
	July	138.5	122.6	2.4			
	Aug.	139.5	125.1	2.5	141.35	2.28	139.92
	Sept.	140.5	127.6	2.5	143.85	2.40	143.20
	Oct.	142.5	129.9	2.3	145.60	2.42	145.73
	Nov.	144.2	132.4	2.5	148.0	2.44	148.26
	Dec.	146.0	134.8	2.4	150.4	2.44	150.66
1976	Jan.	147.9	137.2	2.4	152.8	2.42	152.93
	Feb.	149.8	139.5	2.3	154.45	2.38	154.97
	Mar.	150.6	141.7	2.2	156.0	2.34	156.91
	Apl.	153.5	143.7	2.0	156.7	2.26	158.39

To produce the estimate for May 1976, the moving average of the index must be projected forward 6.5 months, because the latest figure, 143.7, is the underlying figure at the end of October 1975.

Model 4 forecast is thus $143.7 + (6.5 \times 2) = 156.7$

whilst the Model 5 forecast is $143.7 + (6.5 \times 2.26) = 158.39$

Compared with the actual value of 155.2, the predicted values are worse than the previous ones!

We may conclude here that additional complications to a model do not necessarily improve it! It should be noted that these models have been based on a **linear** increase in the index – in practice, the actual increases were more rapid for a time.

Note that the models described here are based solely on the data values given and the trends in such data, and so are examples of **empirical models.** The retail price index is clearly affected by a whole network of government, private and international decisions, and in order to make estimates, for example, a year ahead, a model incorporating these features is needed. The Treasury has a very sophisticated model for predicting future economic trends, but many of the relevant decisions are very difficult to predict. For example, what level of wage increases will be negotiated nationally for the coming year, or what will happen to the foreign exchange rate? So, even using very complicated and detailed models for the economy, predictions must be treated with caution.

7.3 Related problems

1. The values of the R.P.I. are given in the table below from January 1974 until December 1980. Estimate the R.P.I. for the year ahead.

Year	Jan.	Feb.	Mar.	Apr.	May	June	July	Aug.	Sept.	Oct.	Nov.	Dec.
1974	100.0	101.7	102.6	106.1	107.6	108.7	109.7	109.8	111.0	113.2	115.2	116.9
1975	119.9	121.9	124.3	129.1	134.5	137.1	138.5	139.3	140.5	142.5	144.2	146.0
1976	147.9	148.8	150.6	153.5	155.2	156.0	156.3	158.5	160.6	163.5	165.8	168.0
1977	172.4	174.1	175.8	180.3	181.7	183.6	183.8	184.7	185.7	186.5	187.4	188.4
1978	189.5	190.6	191.8	194.6	195.7	197.2	198.1	199.4	200.2	201.1	202.5	204.2
1979	201.2	208.9	210.6	214.2	215.9	219.6	229.1	230.9	233.2	235.6	237.7	239.4
1980	245.3	245.8	252.2	260.8	263.2	265.7	267.9					

(2) The *Financial Times* Index for stock market activity during the first six months of 1980 is given below on a weekly basis. Estimate its value for the next two months (the actual values are given below). What other data might be of some use in estimating how the F.T. index changes?

F.T. Index (based on 1 July 1935 = 100) (1980)

Jan.	4	413.9	Feb.	1	447.8	Mar.	7	455.7	Apr.	3	432.6
	11	435.2		8	461.4		14	439.9		11	435.6
	18	459.8		15	462.6		21	429.9		18	442.7
	25	452.4		22	454.2		28	421.5		25	427.5
				29	467.1						

May	2	443.6	June	6	428.5
	9	436.5		13	456.5
	16	435.7		20	471.8
	23	423.3		27	463.5
	30	415.9			

July	4	483.8	Aug.	1	487.9
	11	493.0		8	481.1
	18	497.3		15	483.8
	25	487.3		22	498.9

8 INVESTMENT OPPORTUNITY

8.1 Problem statement

The sales manager of Tryit Engineering Company has identified a market demand for a new type of product. He is keen for the company to get into this possibly lucrative field, and so has talked the matter over with the production manager. The production manager, however, points out that it would be necessary to buy some new equipment in order to manufacture the product. Together they have worked out estimates for

(i) the purchase cost of the new equipment,
(ii) the lifetime of the machine,
(iii) the running costs of the machine during each year,
(iv) the income from sales of the product during each year.

They have presented their figures to the general manager, along with a recommendation to purchase the machine. Their figures are as follows.

Capital cost			£110 000		
Useful life			5 years		
Scrap value			£10 000 after 5 years		
Cash flows (in £1000 units):					
Year	1	2	3	4	5
Running cost	20	22	24	26	28
Income	56	60	62	60	57
Profit	36	38	38	34	29

Total profit = £175 000 + £10 000 − £110,000
= £75 000.

The general manager, however, is unhappy with this analysis of the situation and has handed the figures over to you, a financial consultant. Develop a mathematical model for use in such areas of decision-making, and apply it to the situation outlined above.

8.2 Possible solutions

The cash flow figures as presented are clearly a great simplification of reality, since they apply only at the end of the year and so ignore the detailed fluctuations within each year. Even allowing for this assumption, however, leaves the figures in error; the analysis as presented has taken no account of interest paid on loans or accrued on savings. For instance, the first year's profit of £36 000 could have been banked for 4 years until the end of the project. With interest compounded annually (which seems a sensible assumption since all the other figures are updated annually) at a rate i, this would thus have increased to £36 000 $(1 + i)^4$. This gives the following amended analysis:

Year	1	2	3	4	5
Profit	36	38	38	34	29
Values at end of project	$36(1+i)^4$	$38(1+i)^3$	$38(1+i)^2$	$34(1+i)$	29

To ensure that we compare like with like, we should also apply this procedure to the original capital cost. Then the total profit at the end of the project is $P = 36(1 + i)^4 + 38(1 + i)^3 + 38(1 + i)^2 + 34(1 + i) + 29 + 10 - 110(1 + i)^5$, which depends crucially on the interest rate i.

Very often a firm has to decide between several alternative investment opportunities not all of which run for the same number of years. Although the above procedure *could* be used to compare two such opportunities, accountants have found it easier to use the idea of NET PRESENT VALUE rather than the value at the end of the project. Thus the profit of £36 000 in one year's time has a net present value of £36 000/(1 + i), since the latter sum invested today would give £36 000 in one year's time. This gives a further analysis.

Year	1	2	3	4	5
Profit	36	38	38	34	29
NPV of Profit	$\frac{36}{1+i}$	$\frac{38}{(1+i)^2}$	$\frac{38}{(1+i)^3}$	$\frac{34}{(1+i)^4}$	$\frac{29}{(1+i)^5}$

with a total profit (in terms of today's money) of

$$\frac{36}{1+i} + \frac{38}{(1+i)^2} + \frac{38}{(1+i)^3} + \frac{34}{(1+i)^4} + \frac{29}{(1+i)^5} + \frac{10}{(1+i)^5} - 110$$

$$= P/(1+i)^5 \text{ from above.}$$

To see how important the value of the interest rate i is for the project value, Fig. 8.1 shows a sketch of P against i.

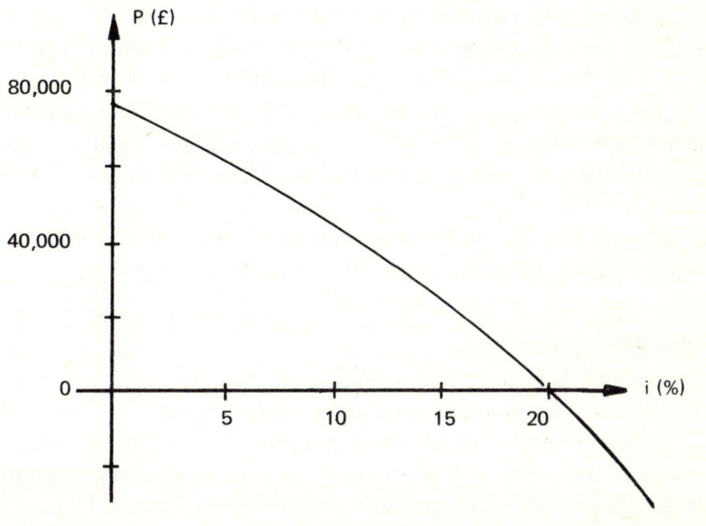

Fig. 8.1

Probably the most convenient way of viewing this graph, is to consider separately the cases where the project value (i.e. NPV) is positive and negative (i.e. $P > 0$ and $P < 0$). If $P > 0$ then you get a greater return from going into the new venture than, for instance, banking the original £110 000 capital. If $P < 0$ then it would be *financially* advantageous to put the whole capital sum into the bank and pay off your workforce, but the unions might not feel this was *socially* advantageous!

In practice an investment opportunity is taken up only if the project value is substantially greater than zero, otherwise an alternative project is searched for with greater profit or less risk.

For the particular project being discussed by Tryit Engineering Company, in a time when interest rates are very close to 20%, great care is clearly needed. The accuracy of the cash flow forecasts would have to be established and the firm's accountants asked about required return on capital. At the end of the day the general manager would have to use his commercial judgement to decide whether to accept or reject this investment opportunity.

9 STOCK CONTROL

9.1 Problem statement

A car distributor holds stocks of oil filters, for each model of which the annual demand is known with a good degree of accuracy. The distributor wishes to develop a stockholding and purchasing policy for these products, to ensure that he is not keeping valuable stock on the shelves for longer than is necessary. He also stipulates, however, that no shortages must occur, since this would lead to dissatisfied customers who might well go to a different outlet in future. He asks you to help him develop this policy and, after some further discussion, you decide to concentrate on one model of filter for the moment. The distributor estimates that the annual demand for this filter will be 10 000 units spread fairly evenly over the year, and he knows that the filters cost £2 each and usually take a month to be delivered from the manufacturers. There is also a cost of £5 incurred each time an order is placed, and an annual cost of holding an item in the store which is estimated to be 18% of the good's price. Can you help him?

9.2 Possible solution

The real problem here is very complicated, since the sales will be subject to a random fluctuation about a trend, as illustrated in Fig. 9.1. Thus at first sight it seems as if we may have to develop a stochastic model to account for this randomness. However, the distributor has told us that the demand is spread fairly evenly over the year, and so it may well be sufficient to assume that the demand is constant. This has the effect of replacing the previous graph with the one shown in Fig. 9.2.

We now have the situation where several different orders (size Q_i) were placed at times T_i, and it would seem sensible to consider the case of constant lot size Q and reorder period T, as shown in Fig. 9.3.

The level S below which the stock is never allowed to fall is called the safety stock, and should ensure that the distributor never runs out of filters even when any randomness is taken into account. We will take $S = 0$ initially, for simplicity, and consider the effect of nonzero safety stock later.

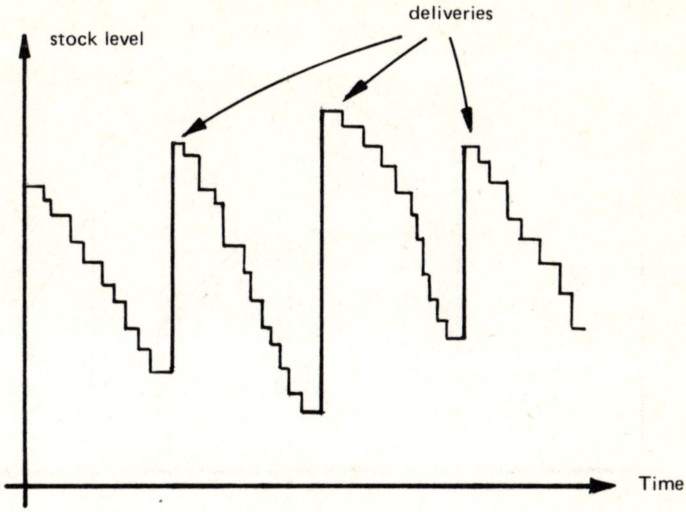
deliveries
stock level
Time

Fig. 9.1

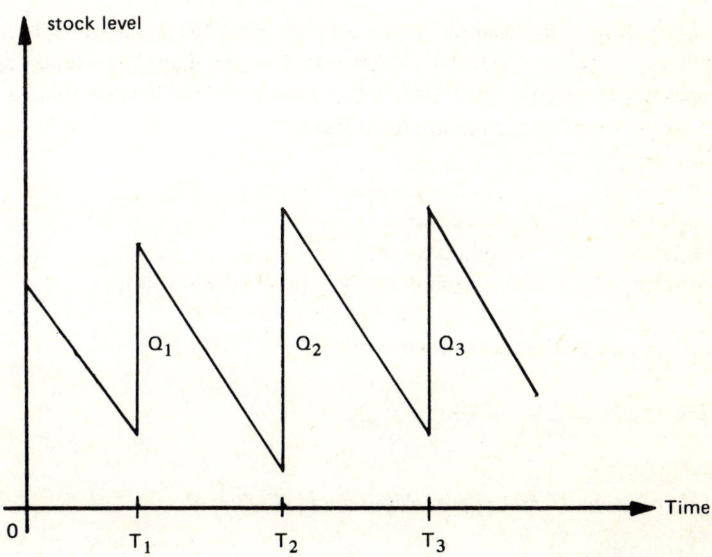
stock level
Q1
Q2
Q3
Time
0
T1
T2
T3

Fig. 9.2

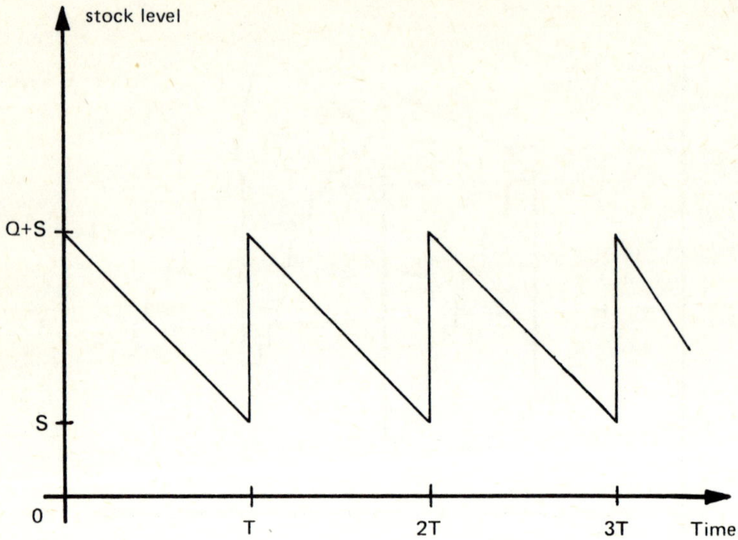

Fig. 9.3

Looking at this idealised graph, we see that the stock level is repeatedly dropping from Q to 0 linearly (in the case $S = 0$). Thus the average stock held throughout the year is ½Q. Also, with a reorder period T years there will be $1/T$ deliveries per year. Introducing the notation

N units/year	=	annual demand
c_1/order	=	order cost
c_2/item	=	purchase cost
a/item	=	stockholding cost (as fraction of price)

enables us to write the total annual cost as

$$C = \frac{1}{T} c_1 + N c_2 + \frac{Q}{2} a c_2 .$$

We also note that T and Q are related via $(1/T)\, Q = N$,
so that

$$C = \frac{N c_1}{Q} + N c_2 + \frac{a c_2 Q}{2}$$

which is sketched in Fig. 9.4.

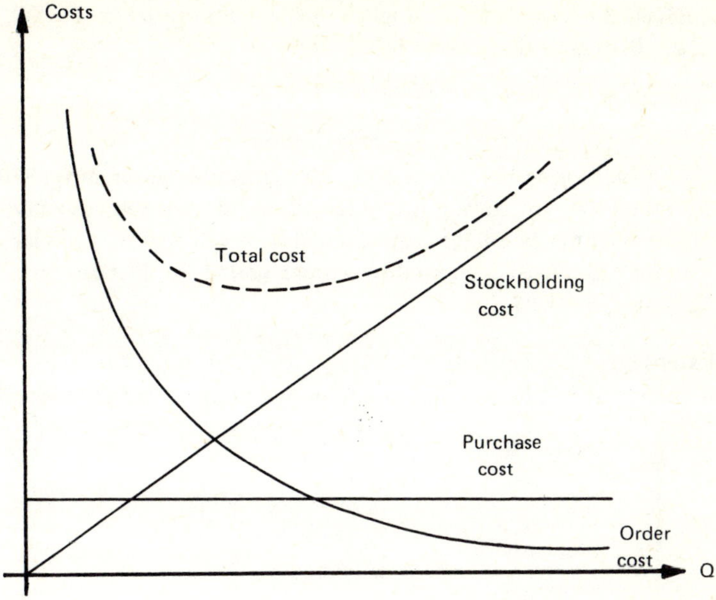

Fig. 9.4

We note that the total cost C has a minimum (which the distributor will wish to know), and that this occurs at

$$Q = \sqrt{\frac{2Nc_1}{ac_2}} \qquad \text{when} \qquad T = \frac{Q}{N} = \sqrt{\frac{2c_1}{aNc_2}} \quad .$$

This formula is known as the ECONOMIC BATCH QUANTITY (or ECONOMIC ORDER QUANTITY) and forms the basis for nearly all the more complex methods of evaluating a stockholding policy.

We now wish to see how the safety stock S affects the EBQ formula. The total cost will be as above, plus the cost of keeping S items in stock all the year. Thus the total annual cost will now be

$$C = \frac{Nc_1}{Q} + Nc_2 + \frac{ac_2Q}{2} + ac_2S$$

which is a *constant* amount greater than before and so has its minimum at exactly the same place. Thus the EBQ formula is unaltered, although the total cost itself is naturally greater.

Applying the formula to the original problem, for which $N = 10\,000, c_1 = 5$, $c_2 = 2, a = 0.18$ gives the optimal figures as

$$Q = 527, \quad T = 0.0527 \approx \tfrac{1}{19} \text{ year}$$
$$\approx 19 \text{ days} .$$

Thus we would probably recommend that the distributor buys 850 filters monthly; the total cost curve is nearly flat around the minimum, so any value of the reorder quantity in a large range around here will lead to a similar cost. In fact, choosing a convenient monthly reorder period involves an extra cost of only £22, or about 0.1%.

9.3 Extension

In many cases it will be possible to obtain discounts for larger order sizes. For

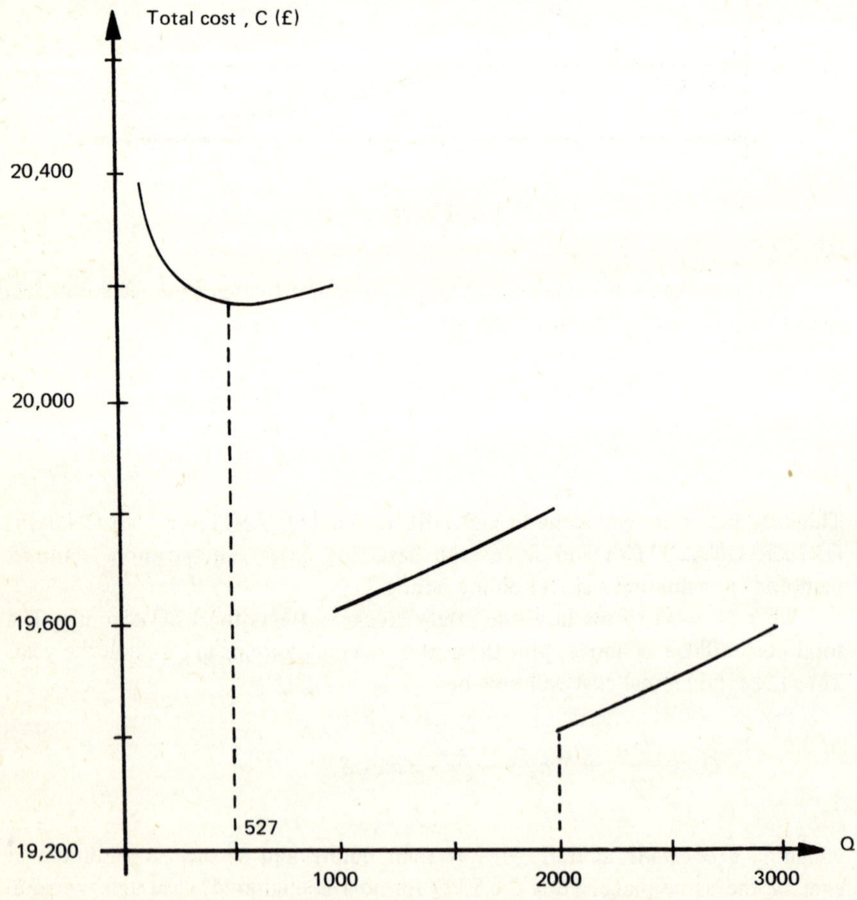

Fig. 9.5

instance, the manufacturers of the filters may charge £2/item for orders of fewer than 1000 items, £1.95/item for orders of between 1000 and 2000, and £1.90/item thereafter. How does this affect the previous analysis?

Although we could still tackle the problem algebraically, it is probably easier to use a graphical approach. In each region where the price c_2 is constant we may use the EBQ formula above. We thus get the following graph (Fig. 9.5), from which it is clear that $Q = 2000$ (and hence 5 orders/year) gives the minimum cost. Again, however, the cost varies little over a large range of Q, and a convenient quantity may be selected without undue financial loss.

10 ADVERTISING EXPENDITURE

10.1 Problem statement

The manager of a chain of DIY stores comes to you, an advertising consultant, with the following problem.

By bulk buying he knows he can obtain a very large stock of white gloss paint at a knockdown price, and he wishes to advertise this in a major drive which the chain of stores is about to undertake. He appreciates that his sales will decrease as he increases the price of the paint (see Table 10.1, which he gives to you) but he needs advice as to the effectiveness of the advertising expenditure. Your previous experience enables you to estimate by how much his sales will increase for a given advertising budget, and you present this (in Table 10.2) via a sales enhancement factor. Thus, for example, by spending £30 000 on advertising you would estimate that his sales would increase by a factor of 1.85.

He tells you, in confidence, that he knows he will be able to get the paint at only £2 per tin, and asks your advice on the optimum advertising expenditure and price. He would also like to know what profit to expect.

Table 10.1 – Expected sales against selling price.

Selling price (£)	Expected sales (thousands of tins)
2.00	41
2.50	38
3.00	34
3.50	32
4.00	29
4.50	28
5.00	25
5.50	22
6.00	20

Table 10.2

Advertising expenditure	Sales enhancement factor
0	1.00
10 000	1.40
20 000	1.70
30 000	1.85
40 000	1.95
50 000	2.00
60 000	1.95
70 000	1.80

10.2 Possible solution

When given data such as these, it is usually a good idea to sketch a graph to see if anything strikes the eye. From Figs. 10.1 and 10.2 it is clear that the data points all lie very close to a well-defined curve – a straight line in the first case and a quadratic in the latter. Thus, introducing the notation

x = expected sales
y = selling price
z = advertising expenditure
c = cost
K = enhancement factor,

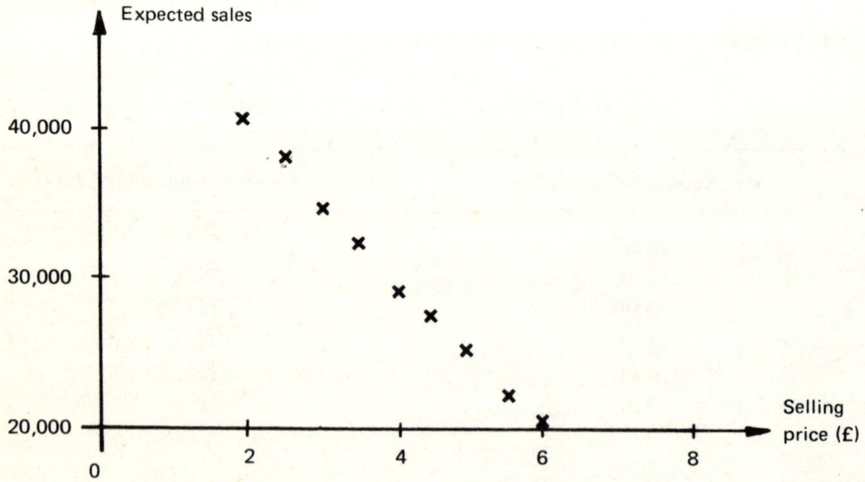

Fig. 10.1

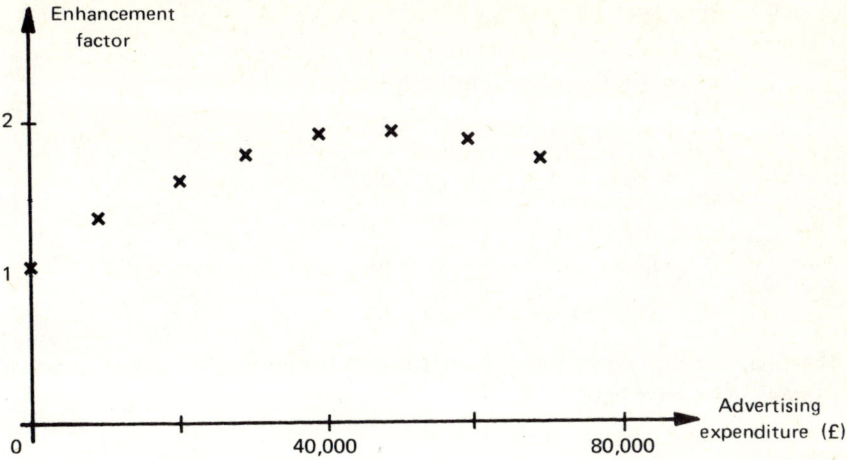

Fig. 10.2

we can say

$$x = a\,y + b$$
$$K = d\,z^2 + e\,z + f$$

(where we expect a and d to be negative). We will, later, require values for these coefficients, but we will derive a general formula first and substitute specific values finally.

Now, the actual sales s are given by multiplying the expected sales by the enhancement factor, so that $s = K\,x$. Then the profit P may be written as

$$\begin{aligned} P &= \text{money in} - \text{money out} \\ &= \text{sales revenue} - \text{purchase cost} - \text{advertising expenditure} \\ &= s\,y - s\,c - z\ , \end{aligned}$$

where we have assumed that all the paint gets sold immediately and so have ignored any short-term cash flow complications. Thus

$$\begin{aligned} P &= s\,(y - c) - z \\ &= K\,x\,(y - c) - z\ . \end{aligned}$$

We also know, however, that $x = x\,(y)$ and $K = K(z)$, so that $P = P\,(y, z)$. We wish to maximise the profit P, and so we need to set $\dfrac{\partial P}{\partial y} = \dfrac{\partial P}{\partial z} = 0$ and solve.

$$P = (dz^2 + ez + f)(ay + b)(y - c) - z$$

$$\frac{\partial P}{\partial y} = (dz^2 + ez + f)(2ay + b - ac)$$

$$\frac{\partial P}{\partial z} = (2dz + e)(ay + b)(y - c) - 1$$

When $\frac{\partial P}{\partial z} = 0$ either $(dz^2 + ez + f) = 0$

or $(2ay + b - ac) = 0$.

The first of these would imply $K = 0$, which is outside the range of practical interest, so we must have

$$2ay + b - ac = 0,$$

i.e. $$y = \frac{ac - b}{2a} .$$

Then $\frac{\partial P}{\partial y} = 0$ gives $z = \frac{1}{2d(ay + b)(y - c)} - \frac{e}{2d}$.

We must now obtain values for all the coefficients. Using the method of least squares gives

$$\begin{aligned} a &= -5133 \\ b &= 50420 \\ c &= 2 \\ d &= -4.256 \times 10^{-10} \\ e &= 4.092 \times 10^{-5} \\ f &= 1.019 , \end{aligned}$$

with a very small residual. [Clearly the curves could also have been fitted by eye, or by choosing representative points and solving the resulting simultaneous equations. The data were only rough estimates and so we would not use highly accurate values for the coefficients. The method of least squares is merely a convenient way of evaluating the coefficients – especially when a computer package is available.]

Substituting these values into the formulae derived above gives

$$\begin{aligned} x &= 20\,084 \\ y &= £5.91 \qquad\qquad K = 1.91 \\ z &= 33\,113 \end{aligned}$$

and so we would forecast that the sales s would be 38 360, with profit $P = £116\,875$.

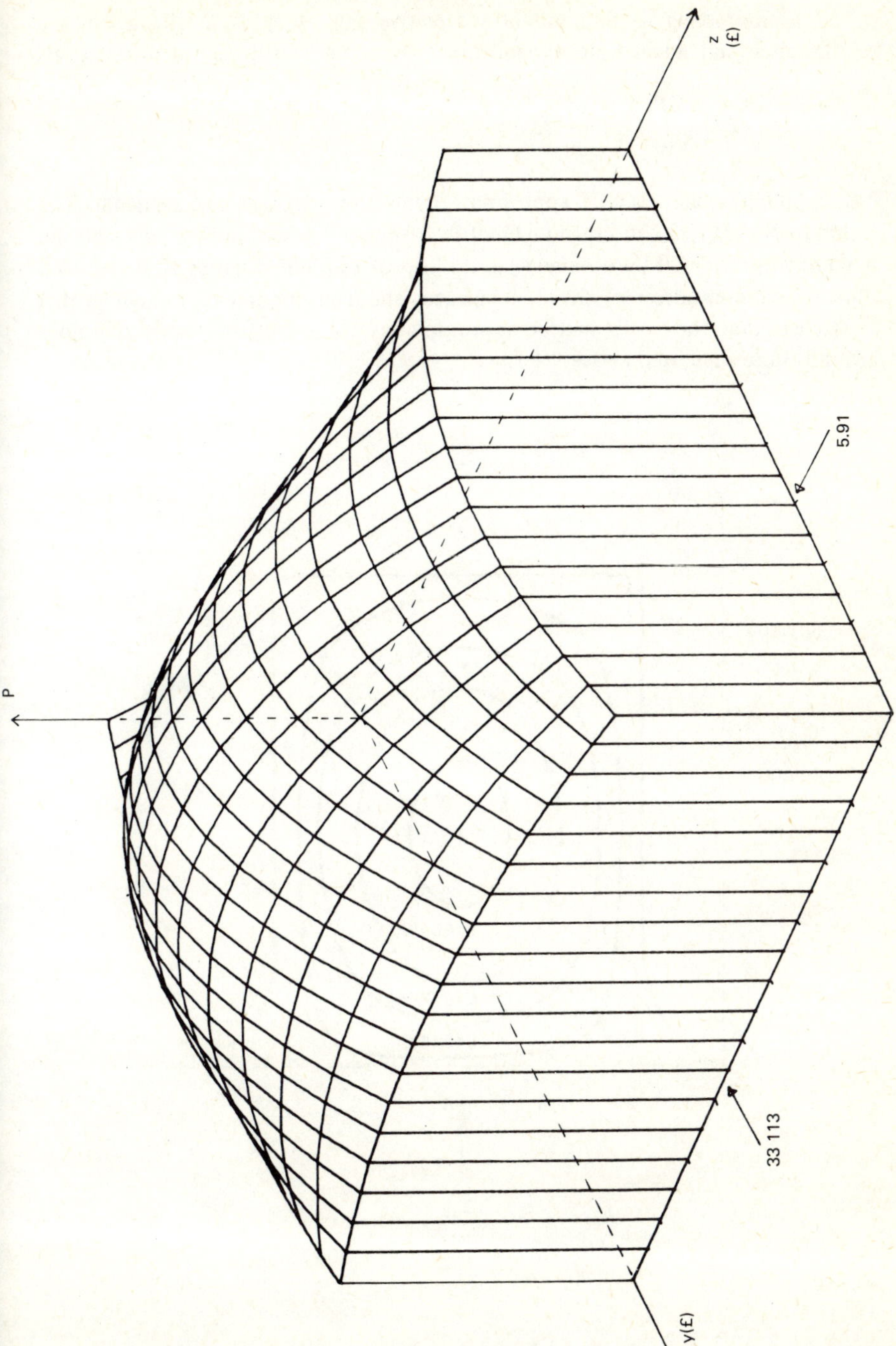

Fig. 10.3

It is interesting to find out how sensitive this profit is to changes in the selling price and advertising expenditure. This could be done via differentials

$$dP = \frac{\partial P}{\partial y}dy + \frac{\partial P}{\partial z}dz \ ,$$

but is much easier seen if appropriate computer packages are available. The function $P = P(y,z)$ can be thought of as a surface in three dimensions, and the maximum profit will be obtained 'at the top of the hill'. Figures 10.3 and 10.4 show a representation of the 3-D surface, and contour lines of equal profit. It is clear that the profit will be approximately the same for a range of values around those suggested above.

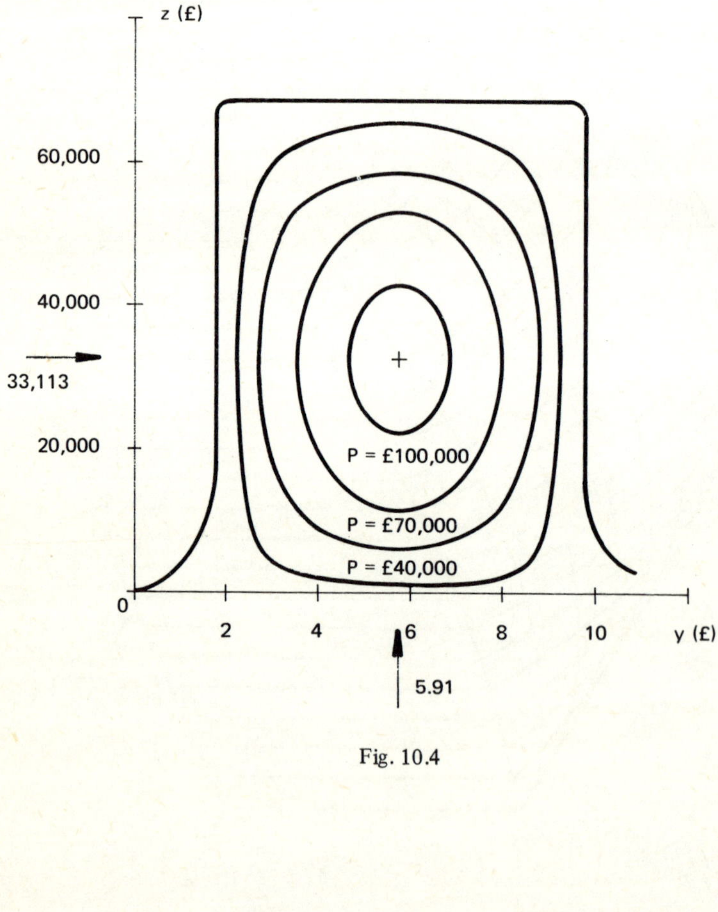

Fig. 10.4

11 COMMUNICATION SATELLITES

1 Problem

It is planned to launch a satellite into a 'stationary' circular orbit above the equator of the earth. In order to remain stationary relative to the earth, what speed should be given to the satellite? Determine the height of the orbit above the earth.

How many such satellites are required above the equator so that all points on the equator are in contact with at least one such satellite?

11.2 Solution

The situation is shown in Fig. 11.1, where plane polar coordinates have been introduced. For this problem, we will use the well-tried and tested Newtonian model for mechanical motion. So the actual modelling content of the problem

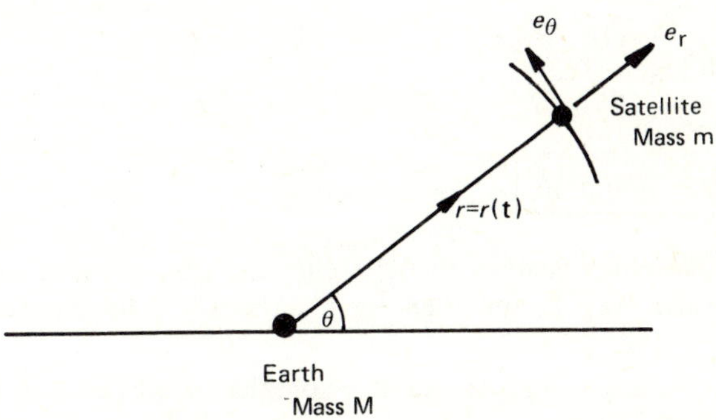

Fig. 11.1 – Satellite Orbit.

is straightforward, but what is important is the use of the model. We will assume that the reader is familiar with planar motion of a particle, using Newton's law of motion and Newton's inverse square law of gravitational attraction. With these assumptions, the equations of motion of the satellite are given by

$$M\ddot{\mathbf{r}} = -\gamma \frac{mM}{r^2}\,\mathbf{e}_r \tag{11.1}$$

where $\mathbf{r} = r\mathbf{e}_r$ is the position vector of the satellite, m its mass, M the mass of the earth, and γ the universal gravitational constant. (The right-hand side of (11.1) represents Newton's inverse square law of gravitational attraction.)

Now, in general, for planar motion using polar coordinates, we have the acceleration

$$\ddot{\mathbf{r}} = (\ddot{r} - r\dot{\theta}^2)\mathbf{e}_r + (2\dot{r}\dot{\theta} + r\ddot{\theta})\mathbf{e}_\theta \tag{11.2}$$

and so we have the equations of motion which can be written as

$$(\ddot{r} - r\dot{\theta}^2)\mathbf{e}_r + (2\dot{r}\dot{\theta} + r\ddot{\theta})\mathbf{e}_\theta = \frac{-\gamma M}{r^2}\,\mathbf{e}_r \quad .$$

Taking components in the $\mathbf{e}_r$ and $\mathbf{e}_\theta$ directions gives

$$\begin{aligned} \ddot{r} - r\dot{\theta}^2 &= -\frac{\gamma M}{r^2} \\ 2\dot{r}\dot{\theta} - r\ddot{\theta} &= 0 \ . \end{aligned} \tag{11.3}$$

The general solution of these equations is in fact a conic, but we can easily see that one particular solution is a circle, radius $r = a$ say, with uniform motion, i.e. $\dot{\theta}$ = constant. The second equation in (11.3) is immediately satisfied (since $\dot{r} = 0$), whilst the first gives

$$a\dot{\theta}^2 = \frac{\gamma M}{a^2} \ ,$$

i.e.

$$\boxed{\dot{\theta} = (\gamma M/a^3)^{1/2}} \quad . \tag{11.4}$$

We can now use this equation to determine the time it takes for a complete orbit of the earth, say T, and then for a stationary orbit we can put $T = 24$ hours.

Now, integration of (11.4) round a complete orbit (i.e. $\theta = 0$ to $\theta = 2\pi$) gives

$$\int_0^{2\pi} \mathrm{d}\theta = \int_0^T (\gamma M/a^3)^{1/2}\,\mathrm{d}t$$

i.e.

$$2\pi = (\gamma M/a^3)^{1/2}\,T \ ,$$

and so

$$T = \frac{2\pi a^{3/2}}{(\gamma M)^{1/2}} \ . \tag{11.5}$$

For a stationary orbit $T = 24$ hours $= 8.64 \times 10^4$ seconds, which gives

$$a = \{8.64 \times 10^4 \times (\gamma M)^{1/2}/2\pi\}^{2/3} \ .$$

Taking $\gamma = 6.67 \times 10^{-11}\ \mathrm{m^3 kg^{-1} s^{-2}}$ and $M = 5.976 \times 10^{24}$ kg gives

$$a = 4.22 \times 10^4\ \mathrm{km} \ .$$

If h = height of the stationary satellite and R = earth's equatorial radius, then

$$h + R = 4.22 \times 10^4 \text{ km}$$

which, using $R = 0.64 \times 10^4$ km, gives

$$\boxed{h = 3.58 \times 10^4 \text{ km}} \quad . \tag{11.6}$$

The satellite's velocity is given by

$$\dot{\mathbf{r}} = a\dot{\theta}\,\mathbf{e}_\theta$$

so that its speed is given by

$$V = a\dot{\theta} = a\left(\frac{\gamma M}{a^3}\right)^{1/2} = \left(\frac{\gamma M}{a}\right)^{1/2}$$

(using 11.4), and with the values above we find that

$$V = 94.45 \text{ km s}^{-1}.$$

To find the required number of satellites needed to give full coverage of all points on the equator, we must firstly determine the angle α, illustrated in Fig. 11.2. Clearly

$$\cos\alpha = \frac{R}{h+R} = \frac{0.64}{4.22}$$

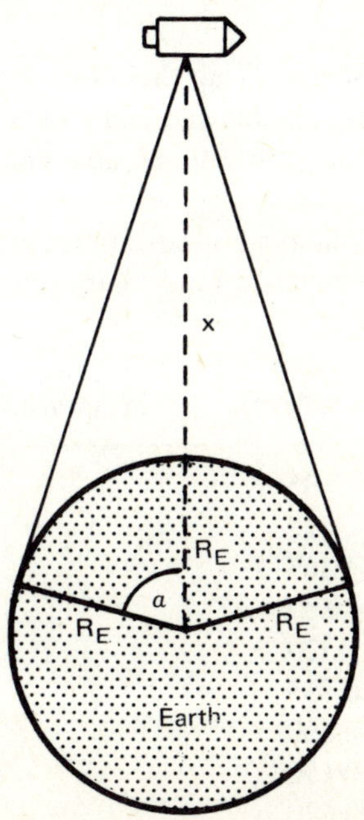

Fig. 11.2 – Communication satellite.

which gives $\alpha = 81°$. So the angle covered by *one* satellite is approximately $162°$, and so *three* satellites would be needed for full coverage of the equator.

11.3 Related Problem

Can communication satellites be placed in 'stationary' orbits about positions *not* on the equator? If so, how many satellites would be needed for complete global coverage? If not, what do you suggest for an efficient communication system?

12 KEPLER'S THIRD PLANETARY LAW

12.1 Problem

In 1601 the German astronomer and science-fiction writer Johann Kepler became director of the Prague Observatory on the unexpected death of Tycho Brahe. Kepler had previously been assistant to Brahe and had helped in collecting 13 years of observations on the relative motion of the planet Mars. By 1609, Kepler had formulated his first two laws:

(i) each planet moves on an ellipse with the sun at one focus,
(ii) for each planet, the line from the sun to the planet sweeps out equal areas in equal times.

Kepler spent the next decade verifying these two laws and formulating a third law, which relates the orbital periods and mean distances from the sun. As with all his laws, this was based on observational data and, published in 1619, it was dedicated to James I of England.

Using the observational data (an up-to-date list is given below), find Kepler's Third Law. Also determine the law from a theoretical model for the solar system.

Planet	Period (days)	Mean distance from sun (km $\times 10^6$)
Mercury	88	57.9
Venus	225	108.2
Earth	365	149.6
Mars	687	227.9
Jupiter	4329	778.3
Saturn	10 753	1427
Uranus	30 660	2870
Neptune	60 150	4497
Pluto	90 670	5907

12.2 Solution: Empirical Model

To determine empirical laws relating variables, we plot firstly the periodic time, T, against the mean distance, R. This is shown in the Fig. 12.1, and it clearly illustrates a nonlinear form for the relationship.

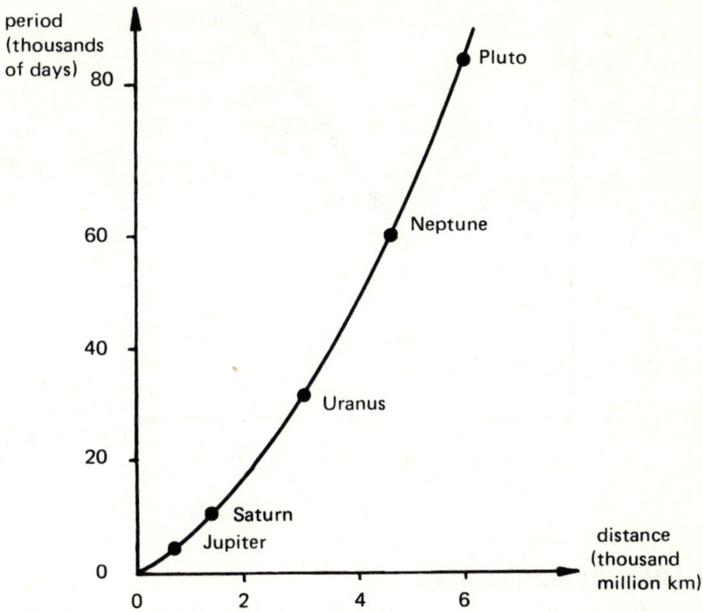

Fig. 12.1

We often look for a power law relationship, say

$$T = KR^{\alpha} , \tag{12.1}$$

in which case

$$\log T = \log K + \alpha \log R . \tag{12.2}$$

So a $\log T - \log R$ plot should be a *straight line* with *slope* α. The graph is shown in Fig. 12.2, and it clearly shows that the data support this hypothesis. The slope is given by $\alpha = 1.5$ and $\log_{10} K = -0.7$ which results in $K = 0.2$. So the data support the model

$$T = 0.2\,R^{3/2} , \tag{12.3}$$

i.e. the square of the orbital time is proportional to the third power of the mean distance. This is in fact Kepler's Third Law.

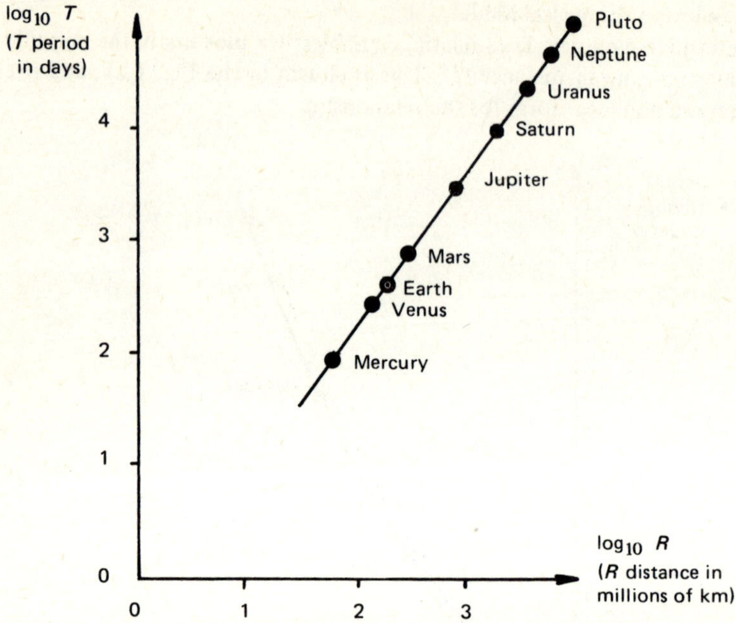

Fig. 12.2

12.3 Solution: Theoretical Model

The theoretical model has already been developed in the section on communication satellites, where equation (11.5) gives the result

$$T = \frac{2\pi R^{3/2}}{(\gamma M)^{1/2}}, \qquad (12.4)$$

where γ is the universal gravitational constant and M the mass of the Sun. Numerical values for these constants are

$$\gamma = 6.67 \times 10^{-11}\ \mathrm{m^3\ kg^{-1}\ s^{-2}}$$

$$M = 1.993 \times 10^{30}\ \mathrm{kg}$$

giving equation (12.4) as $T = 0.199\ R^{3/2}$, which is very close to our empirical model (12.3), above.

12.4 Related Problems

(1) If the solar system were based on an inverse cube law of gravitational attraction, would Kepler's Third Law still be true? (Could planets in fact still move on circular orbits?)

(2) Astronomers in the 17th and 18th centuries observed that the planets of the solar system exhibited some kind of regularity. At that time, the planets Mercury, Venus, Mars, Jupiter, and Saturn were known to exist, and their distances from the Sun had been calculated quite accurately. In 1755 an astronomer, Kant, suggested that there should be another planet between Mars and Jupiter, and in 1766 Titius postulated the relationship

$$R = 0.4 + 0.3\,(2^n)$$

between the planet's mean distance from the Sun, R, and a counter, n, where

n = −1 corresponds to Mercury
n = 0 corresponds to Venus
n = 1 corresponds to Earth
n = 2 corresponds to Mars
n = 4 corresponds to Jupiter
n = 5 corresponds to Saturn
n = 6 corresponds to Uranus
n = 7 corresponds to Neptune
n = 8 corresponds to Pluto

and distances are all measured in Astronomical Units (1AU = average distance of Earth from the sun).

Do the observed data support this proposed law? Can you explain the absence of the number n = 3? (The formula is known as Bode's Law.)

13 A MIRACLE AT MEXICO CITY?

13.1 Introduction

In the 1968 Olympic Games in Mexico City, one result stood out amongst all the others. In his first (and only) attempt at the long jump, Bob Beamon of the USA raised the record from 8.10 m to 8.90 m – an increase of almost 10%. It is still the world record today.

Figure 13.1 shows the world record for the long jump over a number of years, and it is immediately clear what a magnificent feat this was. At the time of writing, Steve Ovett's 1500 m world record is 3 minutes 31.36 seconds; an equivalent performance would cut this record to about 3 minutes 11 seconds!

Several critics have suggested that Beamon's performance was not miraculous at all. They argue that the reduced air density at the high altitude of Mexico City (2256 m) had a great effect on the jump, and the lack of comparable performances by other jumpers was merely because they were not so good on the day. We would like to find out if this uncharitable suggestion has any merit.

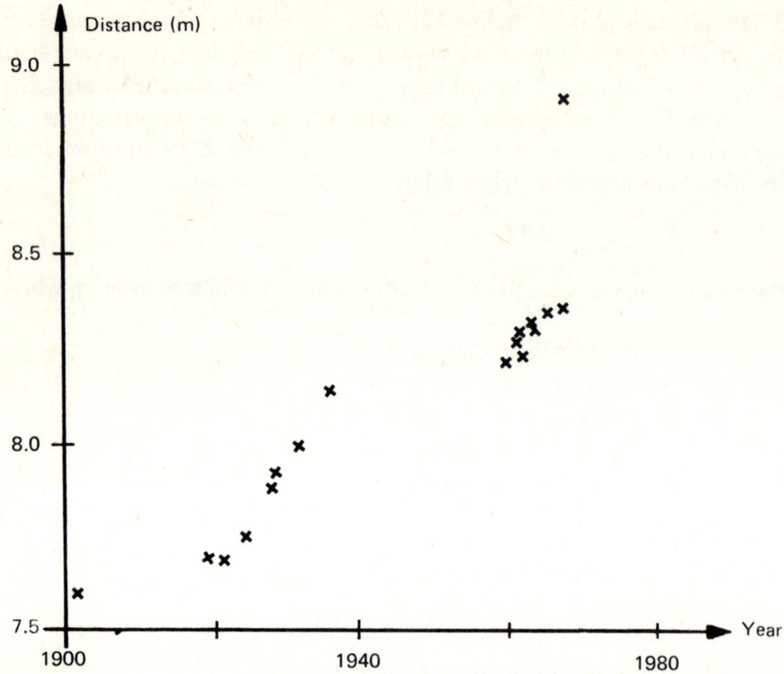

Fig. 13.1

13.2 Possible solutions

The distance which a top-class long jumper can achieve is clearly affected by very many factors. Among these are probably the run-up speed, the spring off the board, the hitchkick, the landing technique, and the athlete's weight and physical shape. To include all of these into our first model is making life unnecessarily hard for ourselves, since a lot of these factors probably have a small effect; anyway, we are not wanting a total model (which takes an athlete and tells us how far he should jump) but an answer to a much more specific question. Thus with any luck we should be able to get by with rather a crude model: by the time we have taken the difference between the long jump distances attained at sea level and at altitude, only small errors ought to remain.

With the above arguments in mind, and knowing that we can always refine our model if the results turn out unrealistic, we will develop a *very* simplistic model. We will assume that the athlete is a projectile fired in a vertical plane, and we wish to compare the ranges of this projectile through media of unknown density. Denoting the jumper's mass by m and velocity by $\mathbf{v}$, Newton's Second Law gives

$$m\frac{\mathrm{d}\mathbf{v}}{\mathrm{d}t} = m\mathbf{g} + \mathbf{D} \ .$$

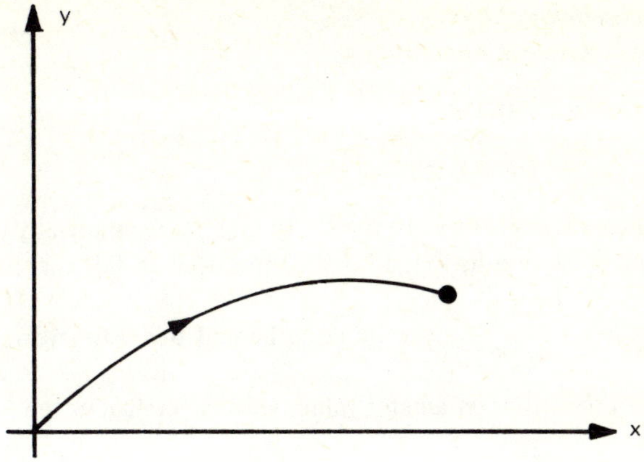

Fig. 13.2

Here $\mathbf{g} = \begin{bmatrix} 0 \\ -g \end{bmatrix}$ is the gravity constant and **D**, a vector in the opposite direction to **v**, is the drag force on the body. For velocities similar to those in this case it is known that

$$|\mathbf{D}| = k\,\rho\,|\mathbf{v}|^2 \ ,$$

where ρ is the air density, and experimentally derived values are available for the constant k. Thus we can write the above equation as

$$\boxed{\frac{\mathrm{d}\mathbf{v}}{\mathrm{d}t} = \mathbf{g} - K|\mathbf{v}|\mathbf{v}} \quad ,$$

where $K = k\rho/m$. To complete the mathematical statement of the problem, we must supply initial conditions for the above nonlinear ordinary differential equation. These are conveniently chosen as

$$\mathbf{x} = 0 \quad \text{and} \quad \mathbf{v} = \mathbf{v}_0 \quad \text{at} \quad t = 0 \ .$$

The above system, however, has no analytical solution. Thus even with this very simple model the solution is non-trivial. We now consider several different methods of obtaining a solution to the nonlinear equation above; in the context of Bob Beamon's jump this is unnecessary, but it provides a useful illustration of the alternatives and allows those without access to a computer to complete the analysis.

13.2.1 *Numerical solution*

The above system can be written as

$$\begin{aligned} \ddot{x} &= -K v \dot{x} \\ \ddot{y} &= -K v \dot{y} - g \end{aligned} \qquad (13.1)$$

and it is then a straightforward matter to integrate numerically using a computer. The trajectory shown in Fig. 13.3 was produced in this manner using an initial velocity $v_0 = \begin{bmatrix} 9.45 \\ 4.15 \end{bmatrix}$; the thinking behind this estimate for the velocity on take-off and the other parameter values used is detailed in the Appendix.

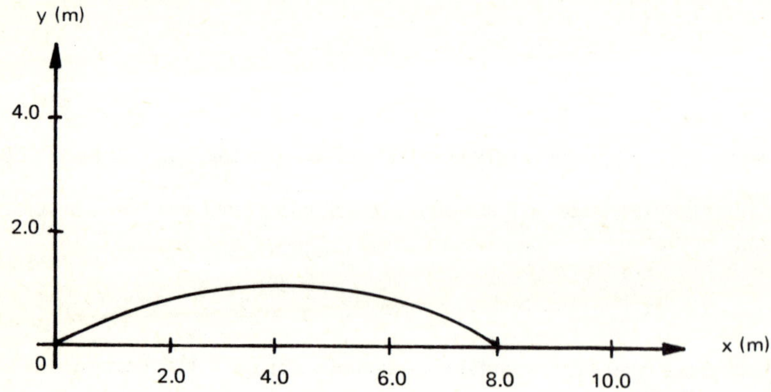

Fig. 13.3

We are now in a position to find out the effect of a change in air density on the range (that is, the value of x when $y = 0$). With the density appropriate to sea level (and other parameters as in the Appendix) the range turns out to be 7.94 m; with the density appropriate to an altitude of 2256 m (Mexico City), the range is 7.96 m. The difference of 2 cm is clearly far too little to explain Beamon's amazing jump!

13.2.2 *Approximate solution*

It would seem obvious that $v = |\mathbf{v}| \approx \dot{x}$ for most of the flight, since the jumper's path is mainly horizontal. Thus we may approximate the system (13.1) by

$$\begin{aligned} \ddot{x} &= -K \dot{x}^2 \\ \ddot{y} &= -K \dot{x} \dot{y} - g \quad . \end{aligned} \qquad (13.2)$$

The first of these equations is easily solved by separation of variables to give

$$\dot{x} = \frac{u_0}{1 + u_0 K t}$$

$$x = \frac{1}{K} \ln (1 + u_0 K t) \ ,$$

and substitution in the second equation and the use of an integrating factor then leads to

$$\dot{y} = \frac{-g}{1 + u_0 K t} \left[t + \frac{(u_0 K t)^2}{2} - \frac{v_0}{g} \right]$$

$$y = \frac{g}{(u_0 K)^2} \left[\left(\frac{1}{2} + \frac{u_0 v_0 K}{g} \right) \ln (1 + u_0 k t) - \frac{u_0 K t}{4} (2 + u_0 K t) \right] \ .$$

Alternatively we can write the solution as

$$y = \frac{g}{(u_0 K)^2} \left[\left(\frac{1}{2} + \frac{u_0 v_0 K}{g} \right) . \ K x - \frac{1}{4} (e^{2Kx} - 1) \right] \ .$$

If this trajectory were drawn on the previous figure, the two curves would be indistinguishable.

To find the range, the value of x at which $y = 0$, we must now solve this equation. Knowing u_0, v_0, and K (see Appendix) allows us to write the equation in the form

$$x = a \left[e^{bx} - 1 \right]$$

which is conveniently solved using the Newton-Raphson algorithm.

Again, changing the air density turns out to lead to only a 2 cm change in the range.

13.2.3 *Energy considerations*

Brearley (1977) puts forward an argument based on the energy dissipated by air resistance, and concludes that the extra range due to a decrease ΔK in K is certainly not greater than

$$\left[1 + \left(\frac{u_0}{v_0} \right)^2 \right] . \left[\frac{2 \, u_0 v_0}{g} \right] . \ \Delta K \ .$$

Using the values given in the Appendix, this turns out to be about 6 cm – again well short of explaining the phenomenal jump.

13.3 Discussion

Having failed to explain the miraculous jump by considering the change in air density, it is natural to look for other factors. We might, for instance, investigate the effect of the reduced gravity at altitude. Remembering that Newton's gravitational law contains an inverse square term allows us to do a rough calculation to see the effect of this change in gravity.

Mean radius of earth	=	6378 km
Altitude of Mexico City	=	2.3 km
Therefore gravity changes by factor		$\left(\frac{6\,378}{6\,380}\right)^2$
	=	0.9994 ,
a decrease of only		0.06% .

Once again the effect is far too small to explain the phenomenal jump. In fact, we are forced to the conclusion that 'everything went just right on the day' for Bob Beamon, and that it was indeed a miraculous jump.

13.4 References

Brearley, M. N., (1977), Article in *Optimal strategies in sport*, edited by S. P. Ladanay, R. E. Machol, North-Holland/American Elsevier.

Hoerner, Y. Y., (1965), *Fluid dynamic drag*, Published by the author.

13.5 Appendix

To estimate the velocity of the long jumper at take-off we may proceed in several ways. The simplest is probably merely to state that he is the equal of an Olympic 100 m sprinter, and make a guess as to the take-off angle. This might give the initial velocity $\mathbf{v}_0 = \begin{bmatrix} u_0 \\ v_0 \end{bmatrix}$ as

$$u_0 = 10 \text{ m/s} ,$$
$$\text{angle} = 25° ,$$
$$\text{therefore } v_0 = 4.7 \text{ m/s} .$$

To improve this initial estimate we might appeal to our knowledge of projectiles *in vacuo*. We know, for instance, that the range is given by $\frac{2\,u_0 v_0}{g}$. Thus if $u_0 = 10$ m/s (perhaps on the high side) and we assume that the jump would be 9 m *in vacuo*, we would obtain $v_0 = 4.4$ m/s.

The quoted values, $u_0 = 9.45$ m/s and $v_0 = 4.15$ m/s, are obtained in this manner. The numerical simulation also uses

$m = 80$ kg
$k = 0.182\ \mathrm{m}^2$ (see Hoerner 1965)
$\rho_{\text{SEA LEVEL}} = 1.225\ \mathrm{kg/m^3}$
$\rho_{\text{MEXICO CITY}} = 0.984\ \mathrm{kg/m^3}$

14 SHOT PUTT ANALYSIS

14.1 Problem Statement

14.1.1 *Background information*

A shot putter concentrates on a smooth build-up of the speed of his body across the circle and extending his arm, thus accelerating the shot in order to give it maximum momentum at the point of release (see pictures overleaf). The shot putter starts in a crouched position with the shot resting under his chin, and finishes in an upright position with arm fully extended. The angle that the extended arm makes with the horizontal will affect the angle of projection of the shot; also the putter 'explodes' into action across the circle, and the speed of projection of the shot depends on this initial burst. The putter thus controls the speed and angle of projection of the shot.

14.1.2 *Problems*

We will not attempt to study all aspects of the shot putter's style by using a simple mathematical model, but there are two aspects of technique which can be investigated using fairly standard methods. Whatever elements of style are used by a sportsman, they are ultimately aimed at projecting the shot at a particular angle of elevation and with maximum initial velocity.

Obvious questions which arise are the effect on the distance the shot is projected of:

(a) the angle of projection
(b) the initial velocity of projection

and

(c) the height of the sportsman.

For some of the problems in this book, it has been difficult to formulate the model, in that it has not been clear what laws connect the variables together, but for this problem we have a well established model, based on Newton's Law of Motion. We will assume that the reader has some previous knowledge of elementary mechanics.

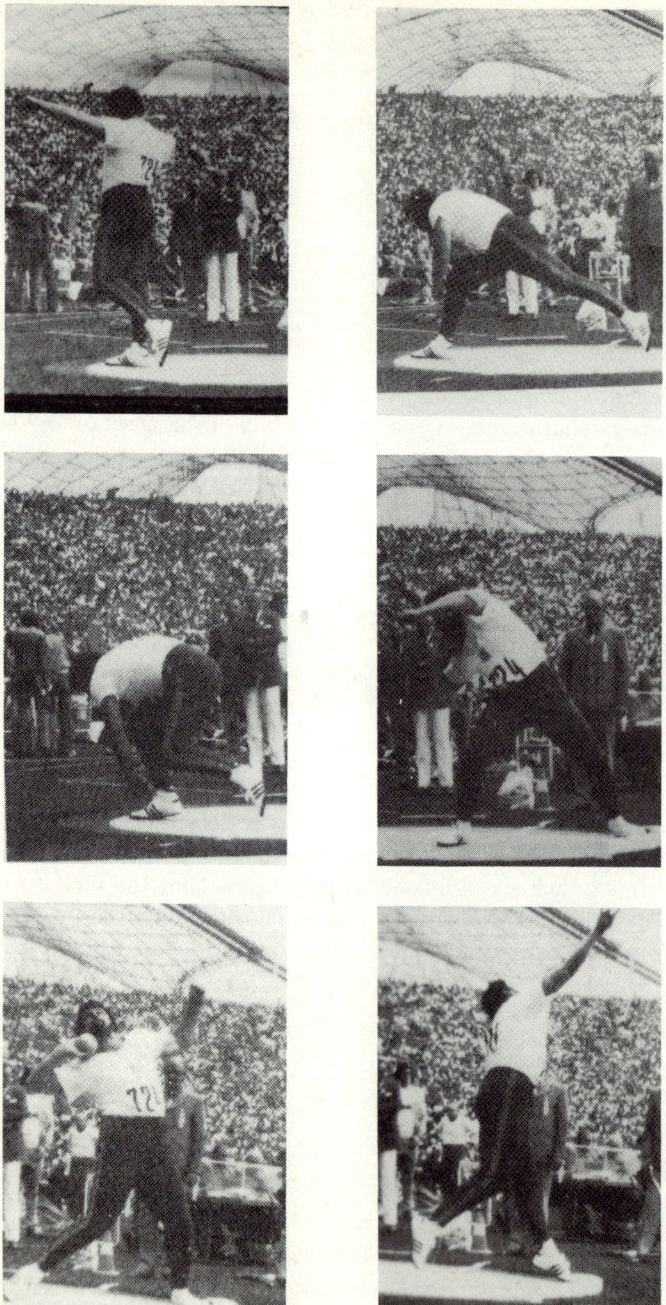
721

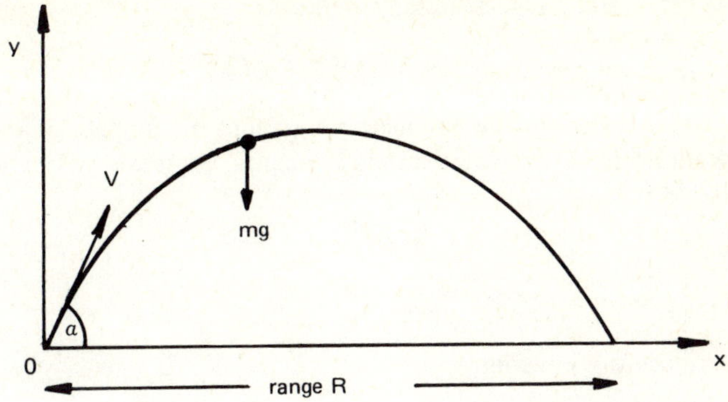

Fig. 14.1

14.2 First model

The simplest model to describe the situation is illustrated in Fig. 14.1. We assume that the shot moves under constant gravity, so that its equations of motion are given by

$$m\,\ddot{\mathbf{r}} = \begin{bmatrix} 0 \\ -mg \end{bmatrix}$$

where m is the mass of the shot, g is the acceleration due to gravity, and $\mathbf{r} = (x,y)$ is the position vector of the shot. In components, the model takes the following form

$$\ddot{x} = 0, \ddot{y} = -g \ . \tag{14.1}$$

This model is completely standard, and a description of its development may be found in any elementary applied mathematics text. Before proceeding, readers should convince themselves that they understand the model and should list the features of the real system which have been set aside in the adoption of this standard model.

Clearly, the horizontal and vertical components of the displacement of the shot can be obtained by integrating equations (14.1) twice. Integrating once gives

$$\dot{x} = A, \quad \dot{y} = -gt + B$$

where A and B are constants. At time $t = 0$, we assume that the shot is projected from ground level, so that we can take $x = y = 0$ and $\dot{x} = V\cos\alpha, \dot{y} = V\sin\alpha$. Hence

$$A = V\cos\alpha \quad \text{and} \quad B = V\sin\alpha \ ,$$

which gives

$$\dot{x} = V\cos\alpha, \quad \dot{y} = V\sin\alpha - gt \ . \tag{14.2}$$

Integrating again, and using the initial conditions $x = y = 0$ at $t = 0$, gives

$$x = (V\cos\alpha)\,t, \quad y = (V\sin\alpha)\,t - \tfrac{1}{2}gt^2 \;. \tag{14.3}$$

Thus we have determined the parametric equations of the path of the particle, and the relationship between vertical and horizontal displacement can be obtained by eliminating t, i.e.

$$y = x\tan\alpha - \frac{gx^2}{2\,V^2\cos^2\alpha} \;. \tag{14.4}$$

We can rearrange this equation as

$$x^2 - \left[\frac{2V^2}{g}\cos\alpha\sin\alpha\right]x = \frac{2V^2}{g}\cos^2\alpha\, y$$

and completing the square yields

$$\left[x - \frac{V^2}{g}\cos\alpha\sin\alpha\right]^2 = \frac{2V^2}{g}(\cos^2\alpha)y + \frac{V^4}{2g}\cos^2\alpha\sin^2\alpha$$

$$\text{i.e.}\quad \left[x - \frac{V^2}{g}\cos\alpha\,\sin\alpha\right]^2 = \frac{2V^2}{g}\cos^2\alpha\left[y - \frac{V^2}{2g}\sin^2\alpha\right] \;. \tag{14.5}$$

Since this is of the form $X^2 = KY$, we have shown that the predicted path of the shot is a *parabola*. This is illustrated in Fig. 14.2.

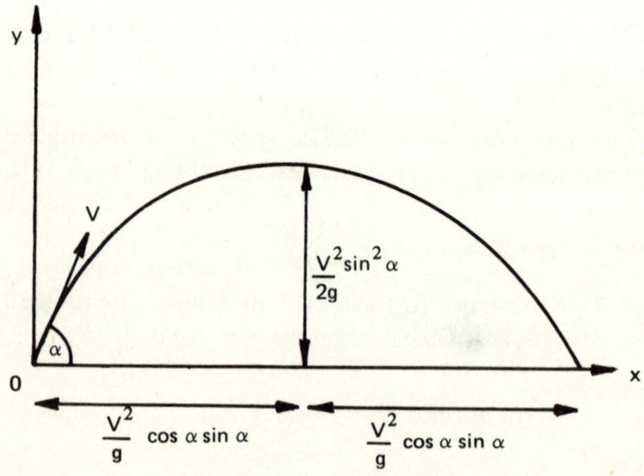

Fig. 14.2 – Parabolic motion.

To determine optimal policies for the shot putter, we must first find the predicted range, R, which is given by the value of x when $y = 0$. From (14.3). $y = 0$ when

$$(V \sin \alpha)\, t = \tfrac{1}{2} g t^2 \ ,$$

i.e. $t = 0$ (which corresponds to the initial projection), and $t = \dfrac{2V}{g} \sin \alpha$. Thus, from (14.3), we have

$$R = (2\, V^2/g) \sin \alpha \cos \alpha \quad ,$$

$$\text{i.e.}\quad R = \frac{V^2}{g} \sin 2\,\alpha \ . \tag{14.6}$$

Clearly, for maximum range, equation (14.6) indicates that we should take $\sin 2\alpha = \pi/2$, giving the optimal angle as $\alpha = \pi/4$. In this case, the range is given by $R = V^2/g$.

Although the model is a very simple one, we can make an attempt to deduce some characteristics of the motion. For example, if the thrower is in error by 10% of the optimal angle of 45°, throwing at an angle of 49.5°, we see that the range, from (14.6), is now given by

$$R = \frac{V^2}{g} \sin 99^\circ \approx 0.99 \frac{V^2}{g} \ .$$

So a *10% error* in the projection angle only results in a *1% decrease* in the range. In other words the range appears to be fairly *insensitive* to the angle of projection. On the other hand, a 5% increase in speed from V to $21\,V/20$ gives a range

$$R = 1.1 \frac{V^2}{g} \sin 2\,\alpha \ .$$

So a *5% increase* in projection speed results in a *10% increase* in the range.

So our first conclusion would be that the major aim for shot putters should be to increase the projection speed, even at the expense of throwing at angles well away from the theoretical optimum value of 45°.

However, is the model a reasonable one? What have we neglected in the model? There are at least three major assumptions which might significantly affect the results.

(i) We have assumed that the shot leaves at ground level, $x = y = 0$, whereas in practice it leaves at $x = 0, y = h$, where h is the height of the thrower. For school competitors this might be very significant, as the total height reached might not be much greater than h!
(ii) We have neglected all forces except gravity, for example air resistance.

(iii) We have assumed that the projection speed does not depend on the angle α. This is certainly not true in practice.

14.3 Improved model

In this model we will try to take into account the criticism (i) above, and so now assume that at $t = 0$

$$x = 0, \; y = h$$

as shown in Fig. 14.3.

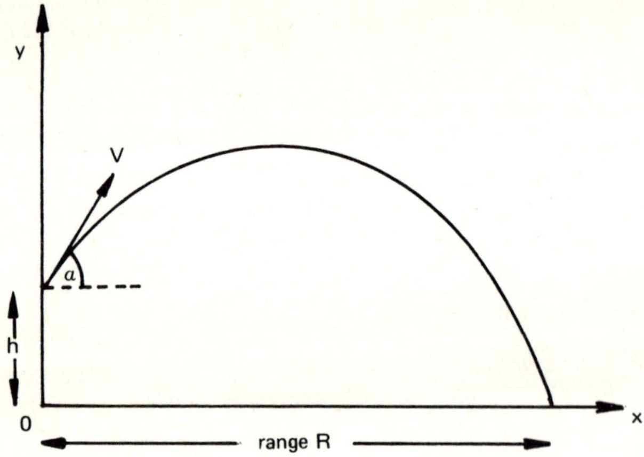

Fig. 14.3 – Improved model for shot putt.

As before, we have equations (14.1) and (14.2), but on integrating (14.2) we obtain

$$x = (V\cos\alpha)t + A, \quad y = (V\sin\alpha)t - \tfrac{1}{2}gt^2 + B$$

where A, B are constants. Since, at $t = 0$, $x = 0$, $y = h$, we have $A = 0$ and $B = h$, giving

$$x = (V\cos\alpha)t, \quad y = h + (V\sin\alpha)t - \tfrac{1}{2}gt^2 \; . \tag{14.7}$$

We still, of course, have parabolic motion, but the formula for the range, R (value of x when $y = 0$), will change. From (14.7), $y = 0$ when

$$h + (V\sin\alpha)t - \tfrac{1}{2}gt^2 = 0$$

i.e. $t^2 - \dfrac{2V}{g}(\sin\alpha)t - \dfrac{2h}{g} = 0$.

Solving this quadratic gives

$$t = \frac{V}{g}\sin\alpha + \left[\frac{V^2\sin^2\alpha}{g^2} + \frac{2h}{g}\right]^{1/2},$$

and so, from (14.7),

$$R = \frac{V^2}{g}\sin\alpha\cos\alpha + \left[\frac{V^2}{g^2}\sin^2\alpha + \frac{2h}{g}\right]^{1/2} V\cos\alpha\ . \tag{14.8}$$

To find the optimum angle of projection, α, we need to calculate $\mathrm{d}R/d\alpha$ and equate to zero. Although this can be done, the algebra becomes very complex, and it is a little easier to note that, from (14.7)

$$R = (V\cos\alpha)t, \quad 0 = h + (V\sin\alpha)t - \tfrac{1}{2}gt^2$$

and, on eliminating t,

$$h + R\tan\alpha = \frac{R^2 g}{2V^2\cos^2\alpha}\ . \tag{14.9}$$

We can now differentiate through this expression with respect to α to give

$$\frac{\mathrm{d}R}{\mathrm{d}\alpha}\tan\alpha + R\sec^2\alpha = \frac{g}{V^2}\left[R\sec^2\alpha\frac{\mathrm{d}R}{\mathrm{d}\alpha} - R^2\sec^2\alpha\tan\alpha\right]\ .$$

Now if $\mathrm{d}R/\mathrm{d}\alpha = 0$, then

$$R\sec^2\alpha = \frac{gR^2}{V^2}\sec^2\alpha\tan\alpha\ ,$$

so
$$R = \frac{V^2}{g\tan\alpha}\ . \tag{14.10}$$

So, for maximum range, this is the optimal relationship between R and α. These variables also satisfy (14.9), and on eliminating R we have

$$h + \frac{V^2}{g} = \frac{V^2}{2g\sin^2\alpha}\ ,$$

i.e.
$$\sin^2\alpha = \frac{V^2/2g}{(h + V^2/g)} = \frac{1}{2(1 + hg/V^2)}\ .$$

Thus the optimal angle α is given by

$$\alpha = \sin^{-1}\frac{1}{[2(1 + hg/V^2)]^{1/2}} \tag{14.11}$$

and, from (14.10), the corresponding maximum range is given by

$$R = \frac{V^2}{g \sin \alpha}(1 - \sin^2\alpha)^{\frac{1}{2}} = \frac{V^2 \left[1 - \frac{1}{2(1 + hg/V^2)}\right]^{\frac{1}{2}}}{g/[2(1 + hg/V^2)]^{\frac{1}{2}}} ,$$

$$\text{i.e. } R = \frac{V^2}{g}(1 + hg/V^2)^{\frac{1}{2}} . \tag{14.12}$$

It should be noted that both (14.11) and (14.12) reduce to the results found in section 14.2 when we put $h = 0$. Let us then consider the significance of the results. Firstly, if we look at the formula (14.11) for α, we see that for $h > 0$, $\sin \alpha < 1/\sqrt{2}$, so that $\alpha < \pi/4$. So the optimal angle is less than the $\pi/4$ value found in section 14.2. To see by how much, we must consider the parameter $\beta = hg/V^2$.

For boys, we can take approximate values,

$$h = 1.5\text{ m} , \quad V = 10\text{ m s}^{-1}, \quad g = 9.81\text{ m s}^{-2} ,$$

which gives $\beta = 0.147$. This is quite small, and using (14.11) we obtain $\sin \alpha = 0.6606$. giving $\alpha = 41°\ 21'$. So the optimum angle is now suggested as about 41°. As in section 14.2, however, we must see how sensitive R is to changes in α or V. From (14.8), we can express R as

$$R = \frac{V^2}{2g}\sin 2\alpha + \frac{V^2}{2g}\sin 2\alpha \left[1 + \frac{2hg}{V^2 \sin^2 \alpha}\right]^{\frac{1}{2}}$$

$$= \frac{V^2}{2g}\sin 2\alpha \left[1 + (1 + 2\beta/\sin^2 \alpha)^{\frac{1}{2}}\right] .$$

Now $\beta/\sin^2 \alpha$ is small in comparison with 1, and so we can approximate R by

$$R = \frac{V^2}{g}\sin 2\alpha \quad (1 + \beta/\sin^2 \alpha).$$

We see again that the dominant term is $V^2 \sin 2\alpha/g$, and so since R is proportional to V^2, it will be more sensitive to changes in V than in α.

So although we have completed a fairly long piece of mathematics, we have arrived at the same conclusion that we found in the simple model in section 14.2. So our original conclusion still holds, but now with even more conviction:

AIM TO INCREASE INITIAL SPEED RATHER THAN
WORRY ABOUT INITIAL PROJECTION ANGLE.

14.4 PROBLEMS

14.4 Problems

(1) Analyse the same problem, but this time include air resistance. For example, suppose that the resistance to motion is proportional either to the speed, V, or to V^2.

(2) Realistic values for h and V, as in section 14.3, are as follows.
(i) For men: $h = 2\text{ m}, V = 15\text{ m s}^{-1}$.

(ii) For women: $h = 1.8\text{ m}, V = 12\text{ m s}^{-1}$.

How do these values affect the conclusions found in section 14.3?

15 POPULATION ESTIMATION

15.1 Problem statement

So that the government can allocate sensibly the money it collects in taxes for building such things as nursery schools and geriatric hospitals, it must have some idea of how many people there will be to use these facilities. Although population censuses are carried out quite frequently, these can only give information about *past* populations.

The tables below show the population of the UK since 1962, together with recent birth and death rates and the age distribution in a typical year. The problem, therefore, is to take these old figures for the UK population and to forecast the population for the next decade.

UK population, births and deaths

Year	*UK population (millions)*	*UK live births (per thousand)*	*UK deaths (per thousand)*
1962	53.274		
1963	53.553		
1964	53.885		
1965	54.218		
1966	54.500		
1967	54.800		
1968	55.049		
1969	55.263		
1970	55.421		
1971	55.610	16.1	12.5
1972	55.781	14.8	12.6
1973	55.913	13.8	12.3
1974	55.922	13.2	11.9
1975	55.900	12.5	11.9
1976	55.886	12.1	12.2
1977	55.852	11.8	11.7
1978	55.836	12.3	11.9
1979	55.883	13.1	—

1975 UK age distribution (millions)

	Males	*Females*
Under 15	6.7	6.4
15-29	6.2	6.0
30-44	5.0	4.9
45-64	6.3	5.0
65-74	2.2	4.6
Over 75	0.9	1.9
Total	27.3	28.8

15.2 Possible solutions

The first thing to do in any problem of this type is to try to understand the data fully. Plotting the population against time gives the graph Fig. 15.1. Thus the population was rising steadily until the early 1970s when, presumably because of the widespread use of the Pill, this rise flattened out.

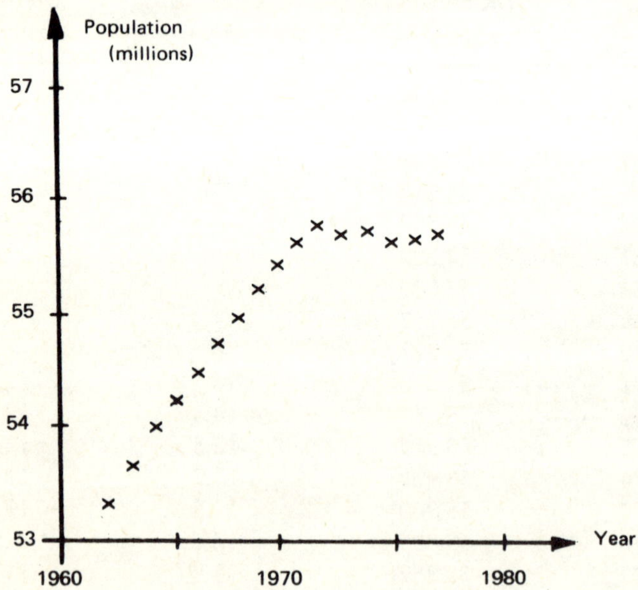

Fig. 15.1

The birth rates given (Fig. 15.2) seem to tell the same story, with the death rate (Fig. 15.3) remaining almost constant.

The data on age structure are much harder to interpret, but may be drawn on a histogram (Fig. 15.4).

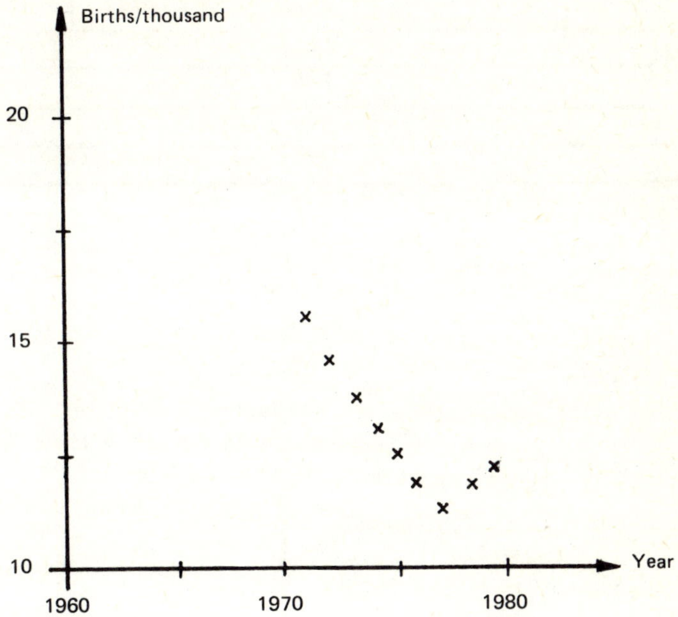

Fig. 15.2

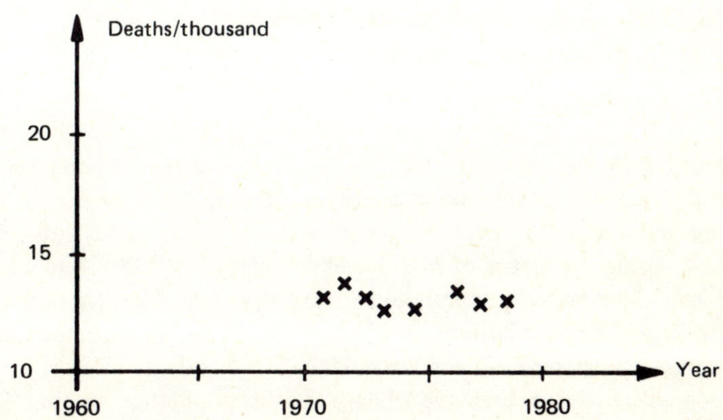

Fig. 15.3

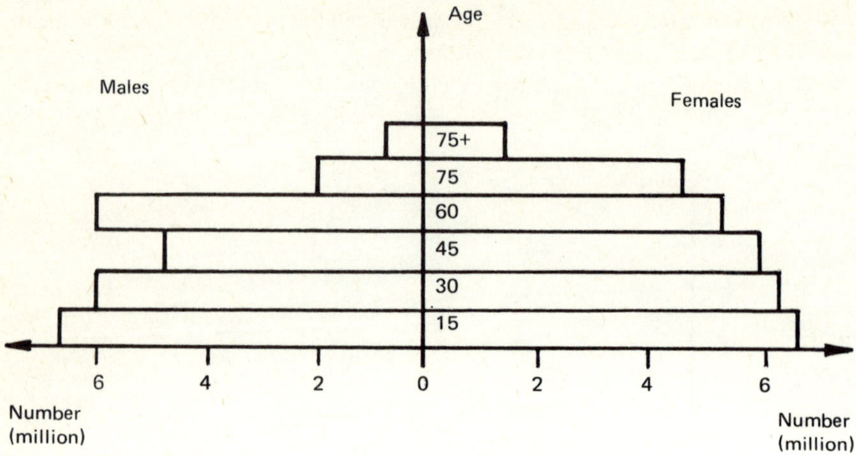

Fig. 15.4

15.2.1 *Model 1*

While the death rate seems to be almost constant, the birth rate exhibits a definite trend. Thus it seems sensible to take the most recent birth rate and the average death rate as estimates for future values. This gives

$$\begin{aligned} \text{birth rate} = b &= 13.1/\text{thousand} \\ &= 0.0131 \ , \\ \text{and death rate} = d &= 12.1/\text{thousand} \\ &= 0.0121 \ . \end{aligned}$$

Letting the population in year t be x_t allows us now to write

$$x_{t+1} = x_t + (b - d)\, x_t \ ,$$

where the last term gives the difference between the *actual number* of births and deaths. This could also be written as

$$x_{t+1} = k\, x_t \ ,$$

where $k = 1 + (b - d)$, and it is clear that $x_t = k^t . x_0$ gives the population in any year t in terms of the population in a base year $t = 0$.

If we define $t = 0$ to be 1962 we have $x_0 = 53.274$, and it only remains to estimate k. Using the values of b and d above gives $k = 1.001$, and we can now use the model to give the population in any year t. In fact, though, it is clear that with $k > 1$ the population $x_t = k^t . x_0$ must always increase; the model with these parameters can never fit the given data.

If we acknowledge that the change in the population in the 1970s was markedly effected by new birth control technology, however, it seems rather harsh to expect our simple model to cope both before *and* after this innovation.

Thus taking $t = 0$ to be 1975 may well be more realistic. This has little effect on our estimates of b and d, and so $k = 1.001$ with $x_0 = 55.9$. We now predict the following behaviour, which is illustrated in Fig. 15.5.

	Year	*Predicted*	*Actual*
0	1975	55.900	55.900
1	1976	55.956	55.886
2	1977	56.012	55.852
3	1978	56.068	55.836
4	1979	56.124	55.883
5	1980	56.180	
6	1981	56.236	
7	1982	56.292	
8	1983	56.349	
9	1984	56.405	
10	1985	56.462	
11	1986	56.518	
12	1987	56.574	
13	1988	56.631	
14	1989	56.688	
15	1990	56.744	

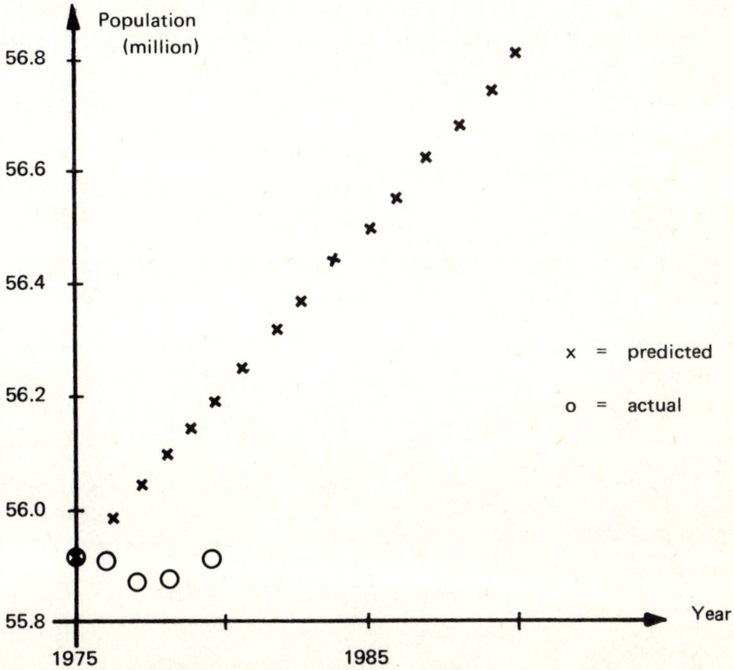

Fig. 15.5

15.2.2 *Model 2*

So far we have not used the data on age structure within the population and have, implicitly, made several assumptions. The two most important of these are that we assumed that births and deaths were proportional to the population size, and that we ignored totally the effects of migration. In fact, the given data *may* be able to give us some information about migration: knowing both birth and death rates for any one year enables us to say *exactly* what the next year's population ought to be – any difference must either be rounding error (and the like) or the net effect of migration. For instance, for 1975 we have

$$x = 55.900 \text{ million}$$
$$b = 0.0125$$
$$d = 0.0119 \; ,$$

and so the 1976 population ought to be 55.934 million instead of 55.886 million. The difference of 48 000 might be due to migration. Unfortunately, repeating this procedure from year to year gives no consistent pattern whatsoever, and we are forced to conclude that such discrepancies are true errors in the data.

To construct a more accurate model, it seems necessary to take some account of the age structure – after all, we can imagine the effects of a population made up entirely of pensioners! Rather than use 6 age classes as in the data, we will try a simple model with only 2 classes. To make it easier to estimate values for b and d we will choose these groups as $0 \rightarrow 14$ and $15+$, and make the assumption that all the births occur in the older group and that 1/20 of the total number of deaths occur in the younger group.

We let $x_1(t)$ and $x_2(t)$ represent the populations of the two groups respectively, and so can write

$$x_1(t+1) = \text{Number of births} + \text{Number staying in group} \cdot \text{Proportion surviving}$$

$$x_2(t+1) = \text{Number changing group} \cdot \text{Proportion surviving} + x_2(t) \cdot \text{Proportion surviving}$$

Denoting the total number of births and deaths by B and D allows us to use the obvious notation

$$B_1 = 0 \qquad D_1 = D/20$$

$$B_2 = B \qquad D_2 = 19D/20 \; .$$

Then we may let

$$x_1(t+1) = b_2 \cdot x_2(t) + \frac{14}{15} x_1(t) \cdot (1 - d_1)$$

$$x_2(t+1) = \frac{1}{15} x_1(t) \cdot (1 - d_1) + x_2(t) \cdot (1 - d_2) \; ,$$

where
$$b_2 = B_2/x_2(t)$$
$$d_1 = D_1/x_1(t)$$
$$d_2 = D_2/x_2(t)$$

and where we have assumed an even age distribution within the younger age group (so that 1/15 of this group become over 15 during the year).

Writing $\underline{x}(t) = \begin{bmatrix} x_1(t) \\ x_2(t) \end{bmatrix}$ allows us to use the concise notation

$$\underline{x}(t+1) = A \,.\, \underline{x}(t)$$

with the transition matrix as

$$A = \begin{bmatrix} \frac{14}{15}(1-d_1) & b_2 \\ \frac{1}{15}(1-d_1) & (1-d_2) \end{bmatrix} .$$

This clearly has solution

$$\underline{x}(t) = A^t \,.\, \underline{x}(0) \quad ,$$

so we now merely require values for the various parameters.

Starting our simulation (as in Model 1) in 1975 (for which we have data on age structure) gives

$$x_1(0) = 13.1$$
$$x_2(0) = 42.9 \quad .$$

We also know that

$$B = 0.0125 \times 55.900 = 0.69875 \text{ million}$$
$$D = 0.0119 \times 55.900 = 0.66521 \text{ million}$$

in 1975, so that

$$b_2 = 0.0163$$
$$d_1 = 0.0025$$
$$d_2 = 0.0147 \quad .$$

We can now predict the population for any year t.

	Year	x_1	x_2	*Predicted total*	*Actual total*
0	1975	13.1	42.9	56.0	55.900
1	1976	12.9	43.1	56.0	55.886
2	1977	12.7	43.4	56.1	55.852
3	1978	12.5	43.6	56.1	55.836
4	1979	12.4	43.8	56.1	55.883
5	1980	12.2	43.9	56.2	
6	1981	12.1	44.1	56.2	
7	1982	12.0	44.3	56.3	
8	1983	11.9	44.4	56.3	
9	1984	11.8	44.6	56.3	
10	1985	11.7	44.7	56.4	
11	1986	11.6	44.8	56.4	
12	1987	11.6	44.9	56.5	
13	1988	11.5	45.0	56.5	
14	1989	11.4	45.1	56.6	
15	1990	11.4	45.2	56.6	

In fact these figures are surprisingly close to those obtained using Model 1, and are shown in Fig. 15.6.

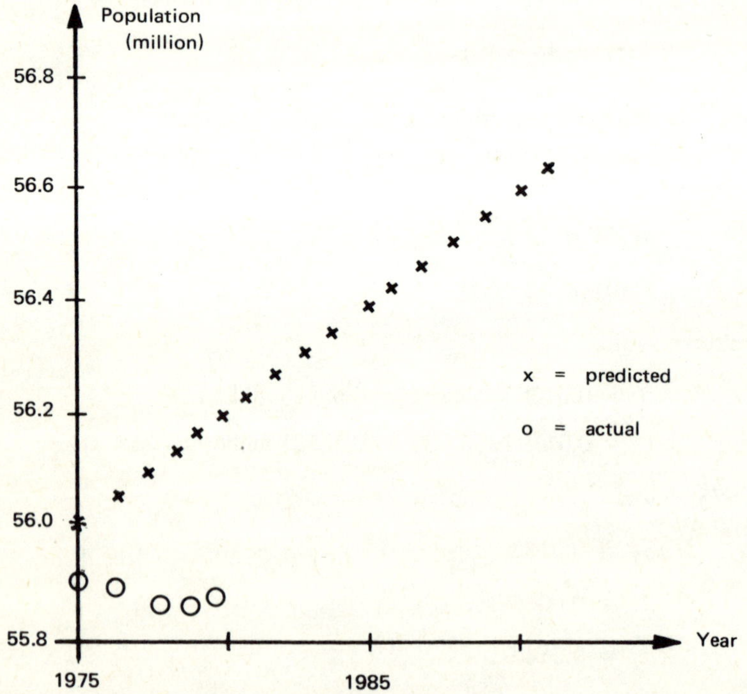

Fig. 15.6

15.2.3 *Comment on Model 2*

It can be shown that Model 2 is, in fact, capable of giving a solution for the total population with the characteristics shown. However, no values of the parameters close to those given above give this form of behaviour.

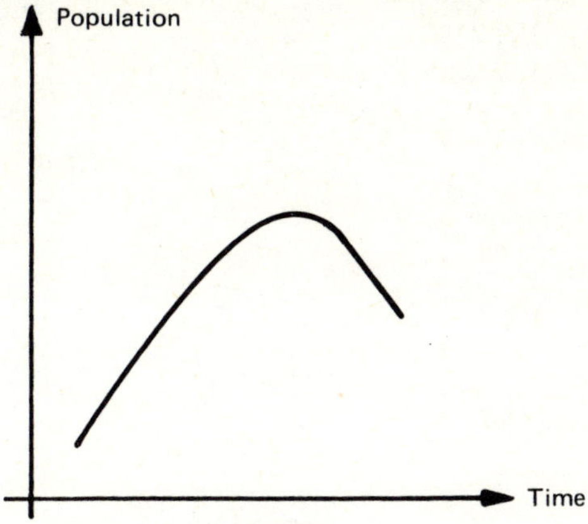

Fig. 15.7

15.2.4 *Model 3*

A commonly quoted example in many books on models is the application of the logistic curve to the USA population, Fig. 15.8.

$$\frac{dN}{dt} = kN \; \frac{A - N}{A}$$

$$N(t) = \frac{A}{1 + e^{a-kt}} \; .$$

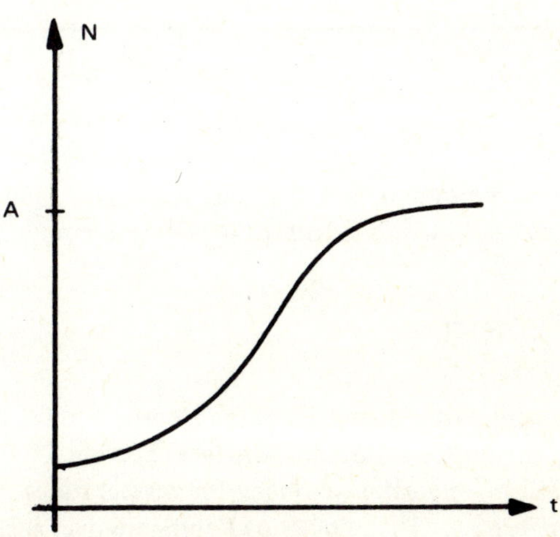

Fig. 15.8

Although there is *absolutely no reason* to expect a fit in the present case readers may be curious as to the result. Using the complete data (i.e. from 1962) and a nonlinear regression package, an excellent fit is obtained with

$$A = 56.253 \text{ million}$$
$$a = -2.832$$
$$k = 0.167/\text{year} ,$$

as shown in Fig. 15.9.

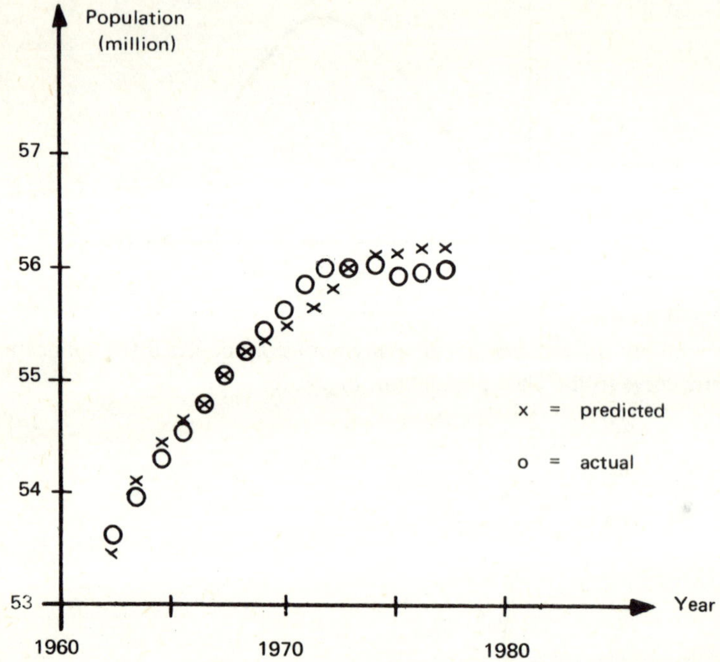

Fig. 15.9

However, it must be emphasised that this is merely duplicating the data and should *not* be confused with modelling the real problem.

16 APPORTIONMENT

16.1 Problem

The apportionment of the members of the House of Representatives in the USA is referred to in the Constitution, Article I, Section 2: "Representatives and direct taxes shall be apportioned among the several states . . . *according to their respective numbers* . . . . The number of Representatives shall not exceed one for every thirty thousand, but each state shall have at least one Representative".

Originally the number of Representatives was set at 65, but Congress does have the power to vary the size of the House. This has steadily increased until, in 1910, it was set at 435, which is the present size of the House.

The constitution does not, though, expressly state what system of apportionment should be used, and in 1881, while considering reapportionment of the members, Congress discovered the following startling fact. Using their current method of appointment, the state of Alabama would be entitled to 8 Representatives in a House having 299 members, but in a House having 300 members it would only receive 7 Representatives! This strange occurrence, known as the 'Alabama Paradox', was in fact no isolated incident.

The method of apportionment being used at this time was called the *Vinton method,* and we will illustrate its use with the following example.

In a college, there are

(i) 103 staff members in the Science Faculty
(ii) 63 staff members in the Arts Faculty
(iii) 34 staff members in the Education Faculty

and the college has 20 members elected to its governing body, Senate, from the three faculties, the number from each faculty being in proportion to the staff size of that faculty.

Faculty	Staff Nos.	% of staff	Exact No. of Senate representatives
Science	103	51.5	10.3
Arts	63	31.5	6.3
Education	34	17.0	3.4
Totals	200	100	20.0

However, representatives of each faculty must be whole numbers! In this way we obtain the following whole representatives:

Science	10
Arts	6
Education	3

giving 19 members. The Vinton method allocates the final seat to the faculty with the highest fractional point – which in this example is the Education Faculty, with 0.4. So we obtain the table

Faculty	% staff	Exact No. in 20-seat Senate	No. of seats
Science	51.5	10.3	10
Arts	31.5	6.3	6
Education	17.0	3.4	4
Totals	100	20.0	20

To illustrate the 'Alabama Paradox', solve the Senate allocation problem described above, but this time with 21 seats to be allocated, and look carefully at what happens to the allocation to Education.

Design a method of apportionment which is more effective than the Vinton method.

16.2 Solutions

For a 21-seat allocation, we have the following table

Faculty	% staff	Exact No. in 21-seat Senate	No. of seats
Science	51.5	10.815	11
Arts	31.5	6.615	7
Education	17.0	3.570	3
Totals	100	21.000	21

After allocating 10 seats to Science, 6 seats to Arts, and 3 seats to Education, the remaining two seats are allocated to the two higher fractional parts, i.e. one to Science and one to Arts, giving the final allocation.

So we have the situation where increasing the number of senate seats from 20 to 21 results in Education losing one seat! We clearly need a better method of allocation.

There are in fact many other ways of allocating seats. We will describe the *'Quota Method'* which is now used in the USA, but you might well arrive at a different but equally effective solution.

We firstly try to quantify the idea of fairness. As a simple example, suppose we have the following representation.

State	Population	No. of Representatives	Population per Representative
A	2 000 000	8	250 000
B	800 000	4	200 000

Since each Representative of state A has to speak for 250 000 constituents whilst each Representative of state B has to speak for 200 000 constituents, it is clear that B is better represented than A. We can take these ideas further by defining, in general,

State	Population	No. of Representatives	
A	p_A	n_A	(*)
B	p_B	n_B	

and the ABSOLUTE UNFAIRNESS of this apportionment as

$$\left| \frac{p_A}{n_A} - \frac{p_B}{n_B} \right| . \tag{16.1}$$

So, from above, absolute unfairness = 250 000 – 200 000 = 50 000.

This measure of unfairness, though, is of little use for comparison purposes. For example, consider the following two examples.

State	Population	No. of Representatives	Pop. per Rep.	Absolute unfairness
C	1542	6	257	32
D	1445	5	289	
X	974 116	4	243 529	64
Y	730 779	3	243 593	

In terms of 'absolute unfairness', the apportionment between X and Y is worse than that between C and D. Our definition of unfairness does not take into account the magnitude of the 'population per representation'; 64 is relatively small in comparison to the constituency size of 243 529, but this is not true of 32 in comparison to 257. So we now define the RELATIVE UNFAIRNESS of an apportionment between states A and B, when state A is more poorly represented than state B, as

$$\frac{\text{Absolute unfairness}}{(\text{population of B/no. of representatives for B})}$$

$$= \frac{p_A/n_A - p_B/n_B}{p_B/n_B} = \frac{p_A}{p_B} \cdot \frac{n_B}{n_A} - 1 . \tag{16.2}$$

For the apportionment between C and D

$$\text{relative unfairness} = \frac{1445}{1542} \cdot \frac{6}{5} - 1 = 0.1245$$

and for X and Y

$$\text{relative unfairness} = \frac{974\,116}{730\,779} \cdot \frac{3}{4} - 1 = 0.0003 .$$

So the representatives have been more equally apportioned between states X and Y than between C and D.

We can use the idea of relative unfairness to assign representatives by

(i) assigning Representative one at a time,
(ii) when assigning a Representative to one of two states, assign it to the state which results in a smaller relative unfairness.

So returning to the situation (*), let us suppose that we now have an extra seat to assign to either A or B. If given to A, we have the situation

State	Population	No. of Representatives
A	p_A	$n_A + 1$
B	p_B	n_B

and the relative unfairness, from (16.2), to state B would be

$$r_B = \frac{p_B}{p_A} \cdot \frac{(n_A + 1)}{n_B} - 1 \; .$$

On the other hand, if given to B we have the situation

State	Population	No. of Representatives
A	p_A	n_A
B	p_B	$n_B + 1$

with relative unfairness to state A being

$$r_A = \frac{p_A}{p_B} \cdot \frac{(n_B + 1)}{n_B} - 1 \; .$$

Now if $r_A < r_B$, then state B receives the extra seat, whereas if $r_A > r_B$ state A gets it. Thus state A gets the next seat when

$$\frac{p_A}{p_B} \cdot \frac{(n_B + 1)}{n_B} > \frac{p_B}{p_A} \cdot \frac{(n_A + 1)}{n_B} \; ,$$

i.e.

$$\frac{p_A{}^2}{n_A\,(n_A + 1)} > \frac{p_B{}^2}{n_B\,(n_B + 1)} \; . \tag{16.3}$$

hence we conclude that the state with the higher value of

$$p^2/n(n + 1)$$

receives the next seat.

We can extend this principle to more than two states. At each stage of the assignment process, the values of

$$Q \equiv p^2/n(n + 1)$$

are computed for every state, and the seat is given to the state with the largest value of Q.

We are now, at last, in a position to solve our original problem. We first assume that each faculty is given *one* seat; and now compute the numbers Q.

Science (1 seat)	*Arts* (1 seat)	*Education* (1 seat)
$\frac{103^2}{1 \times 2} = 5304.5$	$\frac{63^2}{1 \times 2} = 1984.5$	$\frac{34^2}{1 \times 2} = 578$

which shows that the Science Faculty receives the next seat. We now compute the values again

Science (2 seats)	*Arts* (1 seat)	*Education* (1 seat)
$\frac{103^2}{2 \times 3} = 1768.2$	$\frac{63^2}{1 \times 2} = 1984.5$	$\frac{34^2}{1 \times 2} = 578$

and so Arts Faculty gets the next seat, and we continue

Science (2 seats)	*Arts* (2 seats)	*Education* (1 seat)
$\frac{103^2}{2 \times 3} = 1768.2$	$\frac{63^2}{2 \times 3} = 661.5$	$\frac{34^2}{1 \times 2} = 578$

and this seat goes to the Science Faculty.

By now, it is clear that at each assignment we repeat the computation of 2 out of 3 values. After a little thought, it is also clear that if we compute the values

Science	*Arts*	*Education*
$103^2/(1 \times 2)$	$63^2/(1 \times 2)$	$34^2/(1 \times 2)$
$103^2/(2 \times 3)$	$63^2/(2 \times 3)$	$34^2/(2 \times 3)$
⋮	⋮	⋮

then the first 20 seats are allocated to the 20 highest values. The actual values are given in the table below.

Science		*Arts*		*Education*	
5304.5	4.	1984.5	5.	57.8	9.
1768.2	6.	661.5	8.	192.7	15.
884.1	7.	330.8	12.	96.3	
530.5	10.	198.5	14.	57.8	
353.6	11.	132.3	18.	38.5	
252.6	13.	94.5			
189.4	16.	70.9			
147.3	17.	55.1			
117.9	19.				
96.4	20.				
80.4					

The allocation of seats 4-20 is shown, and so we have the allocation of 20 seats as

Science	Arts	Education
11	6	3

and from the table we see that the 21st seat will be allocated to Education.

16.3 Related Problems

(1) Ten seats on the halls of residence committee are to be apportioned to three halls A, B, and C. There are 235 students in A, 333 in B, and 432 in C. Use an apportionment method to allocate the seats.

(2) If the representation on the committee is increased to 15, how should the 5 new seats be distributed?

(3) Does the apportionment principle described above give an ideal method for apportioning members in the House of Representatives in accordance with the constitution?

17 FORESTRY MANAGEMENT

17.1 Introduction

The forestry manager faces many interesting biological and economic problems, and we here consider the problem of forest rotation. The commercial value of a single tree is determined by the volume and the quality of timber which the tree can produce, but clearly also depends on the age of the tree. Figure 17.1 shows this dependence in the case of British Columbia Douglas Fir (from Clark 1976). It is recognisable that young trees have no commercial value, and that the value increases with age as the volume of usable timber increases. Eventually the tree approaches maturity, its growth ceases, and its value reaches a plateau. Ultimately decay would set in and the value decline to zero.

Although the details of this curve depend on the species of tree and the use to which the timber is to be put, the general shape is common to all commercial trees. The problem for the forestry manager, then, is to use such curves to decide when to fell a stand of trees of the same age.

17.2 First thoughts

Assuming that the value $V(t)$ is known and c denotes the cost of felling, $V(t) - c$ represents the net value of the stand of trees (the net stumpage value, as in the figure). Thus it would seem obvious that we merely choose the value of t which makes $V(t) - c$ maximum, so that for Douglas Fir we harvest when the trees are about 120 years old. This, however, ignores the fact that a given amount of cash we have on hand now is more valuable to us than the same amount received at some later date. [This is true even without inflation, since money on hand may be invested to produce more money.]

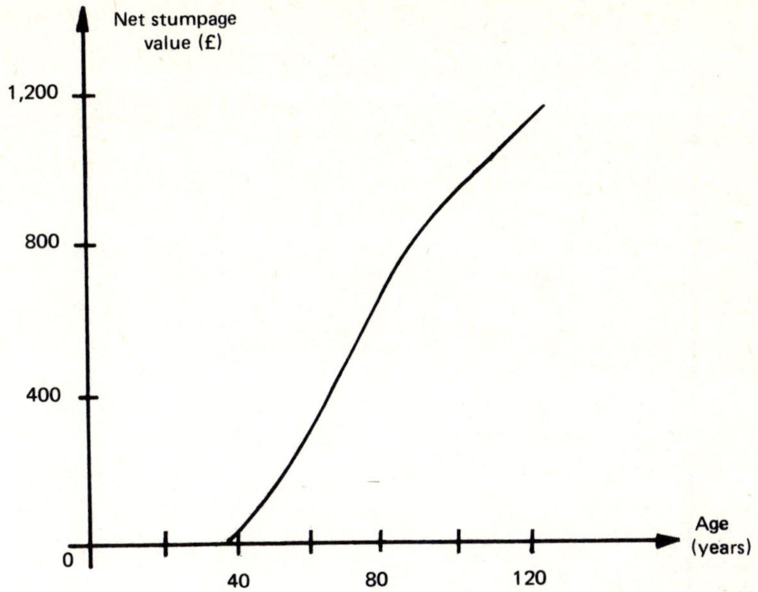

Fig. 17.1

To account for this we discount the stumpage value by the factor e^{-rt}, where r is the continuous interest rate, and obtain the present value of the future earnings as

$$PV = e^{-rt}\,[V(t) - c] \; .$$

To obtain the year $t = T$ at which the present value will be maximum we differentiate this expression and obtain

$$\frac{V'(T)}{V(T) - c} = r \; .$$

The Appendix contains the data concerning $V(t) - c$ from which Fig. 17.1 was drawn, and so it is a simple matter to use a central difference formula to approximate $V'(t)$:

$$V'(t_n) = \frac{[V(t_{n+1}) - c] - [V(t_{n-1}) - c]}{t_{n+1} - t_{n-1}} \; .$$

Figure 17.2 shows the expression $\dfrac{V'(t)}{V(t) - c}$, known as the relative biological growth rate, plotted against time t.

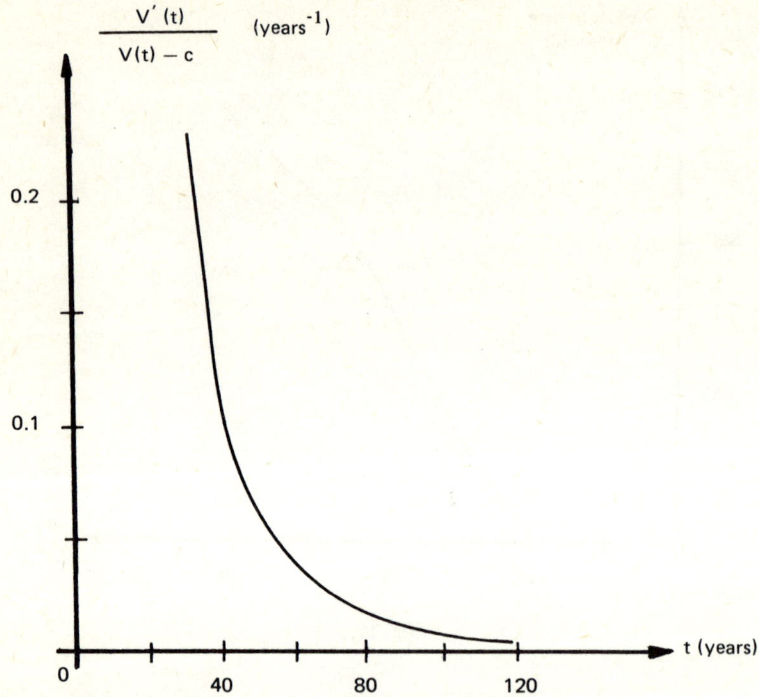

Fig. 17.2

Knowing the prevailing interest rate r, we can now read off the age T at which the trees should be felled. For instance, if $r = 10\% = 0.1$ we would choose to harvest Douglas Fir at an age of about 50 years.

17.3 Further thoughts

The solution above ignores an important aspect of the forest rotation problem: once the trees are removed from a given area, the land is available for new forest growth. Clearly, the longer the felling of the existing forest is delayed, the longer it takes to acquire revenue from future harvests. We also notice that after each successive harvest we are faced with a problem identical to the original one: when should the next crop of timber be cut? Thus it follows that if r stays constant each crop should be allowed to grow for the same time T. Letting c now represent the sum of felling *and* replanting, we can generalise our original solution to obtain the total present value of all future harvests as

$$PV = e^{-rT}[V(T)-c] + e^{-2rT}[V(2T-T)-c] + e^{-3rT}[V(3T-2T)-c] + \cdots$$

$$\text{So } PV = \sum_{n=1}^{\infty} e^{-nrt}\,[V(T)-c]$$

$$= \frac{V(T)-c}{e^{rT}-1}\ ,$$ where we assume an infinite time horizon. Maximising this expression with respect to T then requires that

$$\frac{V'(T)}{V(T)-c} = \frac{r}{1-e^{-rT}}\ ,$$

which is known as the Faustmann formula after a 19th-century German forester.

To obtain the optimal rotation period T, we must solve the above equation. The left-hand side we have seen before: it is the relative biological growth rate sketched in Fig. 17.2. Thus if we graph the right-hand side (known as the relative economic growth rate) too, and superimpose the two, we may read off the value of T for any interest rate r. This is shown in Fig. 17.3. It is clear that, since Douglas Fir takes a considerable time to mature, the effect of crop rotation is very small.

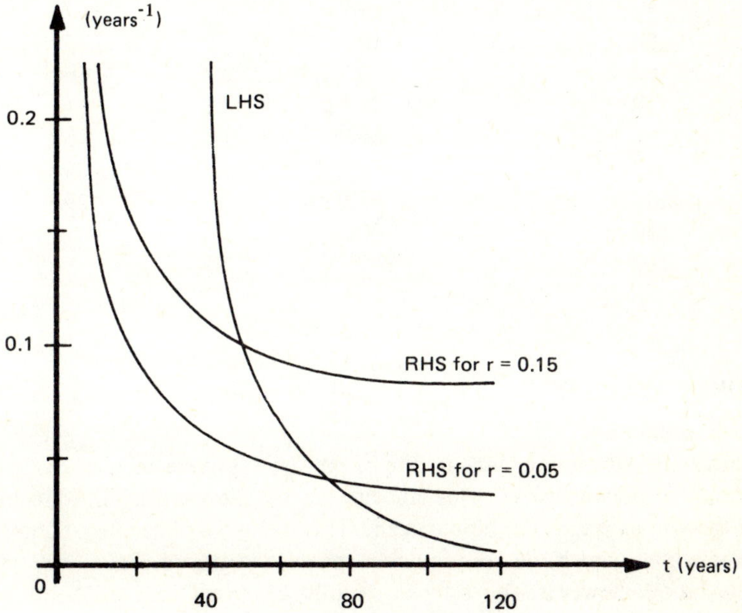

Fig. 17.3

The Appendix also contains figures for European Larch, however, a tree which matures much earlier; a similar analysis then shows a much more marked effect. Graphs such as these are of great value to forestry management.

17.4 Parting comments

(1) It could well be claimed that the manager ought to be maximising the average annual yield $(V(T) - c)/T$. Confirm that this corresponds to zero discounting $(r = 0)$.

(2) The Faustmann model can be used to predict the response of the forest industry to a change in the demand for forest products or in the relative cost of logging. For further details see Clark (1976).

17.5 References

Clark, C. W., (1976), *Mathematical Bioeconomics,* Wiley Interscience.
Pearse, P., (1967), The Optimal Forest Rotation, *Forestry Chronicle,* **43**, 178–195.

17.6 Appendix

Typical net stumpage values, $V(t) - c$, for Douglas Fir and European Larch (from Pearse 1967).

Age (years)	*Fir value (£)*	*Larch value (£)*
20	0	0
30	0	60
40	43	132
50	143	198
60	303	258
70	497	319
80	650	368
90	805	405
100	913	429
110	1000	
120	1075	

18 DRUG THERAPY

18.1 Introduction

Patients with asthma have constriction of the airways in the lungs and consequent difficulty in breathing out. This ailment can be alleviated by introducing the drug theophylline into the bloodstream. This is done by injecting another drug, aminophylline, which the body quickly converts to theophylline. Once present in the blood, however, the drug is steadily excreted from the body via the kidneys. In other words, the system 'leaks', and unless there is replenishment the quantity of drug in the blood will fall.

It is known from experiments that theophylline has hardly any therapeutic effect if its concentration in the bloodstream is below 5 mg/l, and that concentrations of above 20 mg/l are likely to be toxic. The problem is to administer the drug in such a way that the concentration remains within the therapeutic range between 5 mg/l and 20 mg/l.

Experimental measurements can be obtained by injecting a known dose of theophylline into a patient, allowing time for it to diffuse throughout the bloodstream, and finding the drug concentration in blood samples taken at regular intervals. One then knows the initial quantity put into the bloodstream and the concentration at various later times, but there is no time for which both the quantity and the concentration are known. The following data come from such an experiment.

Initial quantity = 300 mg

Concentration, mg/l	Time, hours
10.0	1
7.0	3
5.0	5
3.5	7
2.5	9
2.0	11
1.5	13
1.0	15
0.7	17
0.5	19

Another experimental observation is that the apparent volume of distribution (V litres) and the patient's weight (W kg) are connected by the simple relationship $V = \frac{1}{2}W$. Thus the dose necessary to achieve a required initial theophylline concentration can be inferred from the patient's weight alone: the dose D mg to obtain a 12 mg/l concentration in a 50 kg patient is obtained from

$$D/V = 12 \ ,$$

so that $$D = 12\ V = 6\ W = 300 \text{ mg.}$$

Assuming that the rate at which the drug is removed by the kidneys is proportional to the amount of drug in the body, obtain a differential equation which describes the concentration $c(t)$ at any time t. Solve this equation, and apply the answer to the data given above to see how well the model agrees with reality.

It is clear that in order that the drug concentration remains inside the therapeutic range a series of injections must be given, and the desired concentration pattern may be as in Fig. 18.1.

By considering a very simple case where equal sized doses D mg are administered at equal intervals of time T hours, confirm that it is possible to achieve a saturation as in the Fig. 18.1. Is it possible to choose the dose D and the interval T to remain in the therapeutic range? What would you advise for the patient mentioned above? Is there any advantage in giving a larger dose initially and then a series of doses of size D as above?

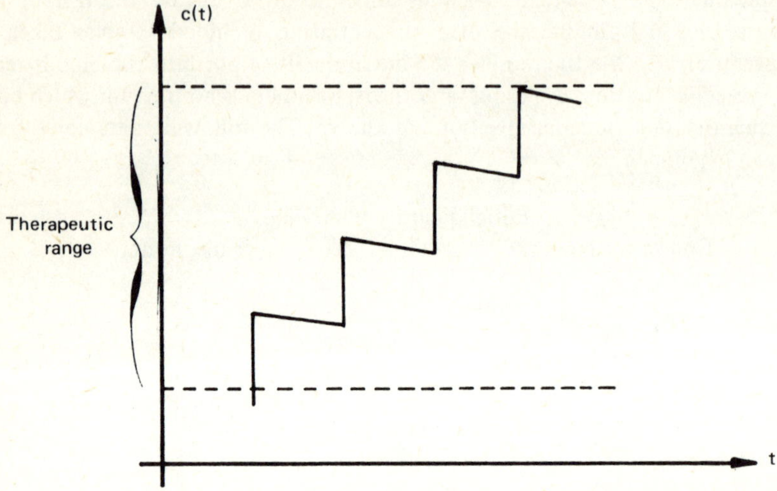

Fig. 18.1

18.2 Possible solution

When given a question involving data, it is always a good idea to plot them. Doing so in this case gives the graph shown in Fig. 18.2. This reminds us of an exponential curve, and replotting as ln c against t (Fig. 18.3) confirms this guess.

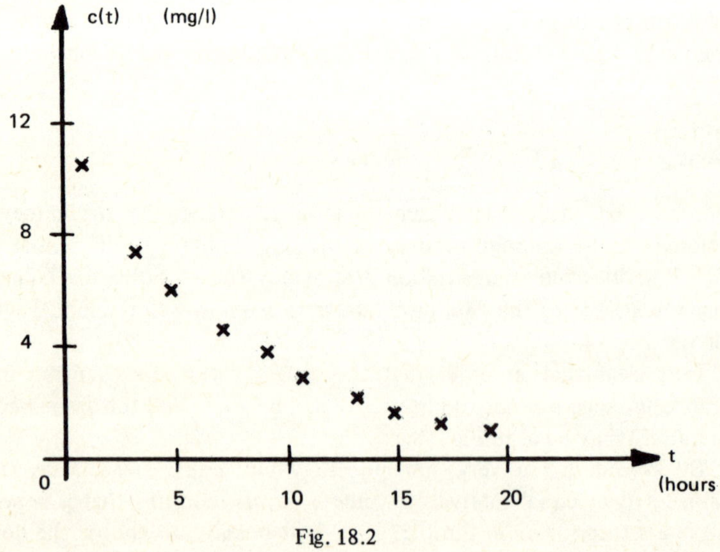

Fig. 18.2

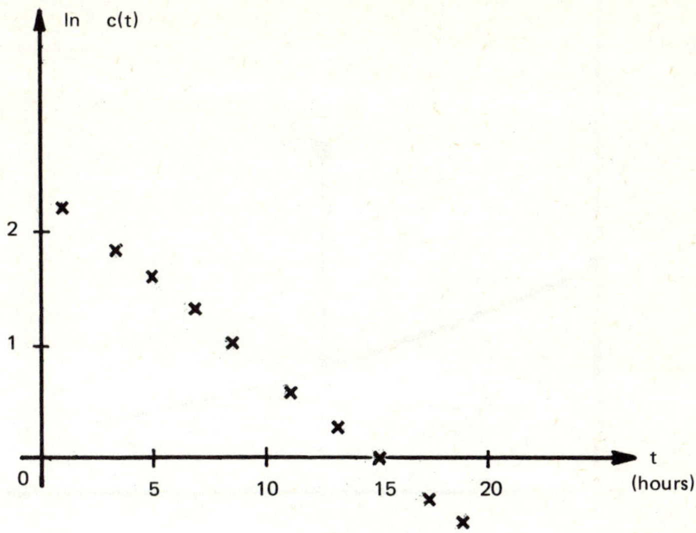

Fig. 18.3

Thus we expect that our theory will yield something like $c(t) = c_0\, e^{-kt}$. To confirm this result theoretically we note that the drug is removed by the kidneys at a rate proportional to the amount present. Thus if $y(t)$ is the amount of drug present at time t, then

$$\mathrm{d}y/\mathrm{d}t = -ky \ .$$

We also know, though, that the amount of drug and the concentration are connected via the apparent volume:

$$c(t) = y(t)/V \ .$$

Thus $\mathrm{d}c/\mathrm{d}t = -kc$, giving us

$$c(t) = c_0 e^{-kt} \text{ as anticipated.}$$

To complete this part of the analysis, we must evaluate the parameters c_0 and k. The linear graph, Fig. 18.3, is ln c against t. However, we have

$$\ln c = \ln c_0 - kt \ ,$$

so the intercept on the graph is ln c_0 and the slope is $-k$. This gives

$$c_0 = 12 \text{ mg/l}, \quad k = 0.17/\text{hour}$$

and agrees well with the data given for the 50 kg patient.

We now assume that another dose of D mg (where $D/V = c_0$) is given at time T. This will yield the form of concentration with time shown in Fig. 18.4,

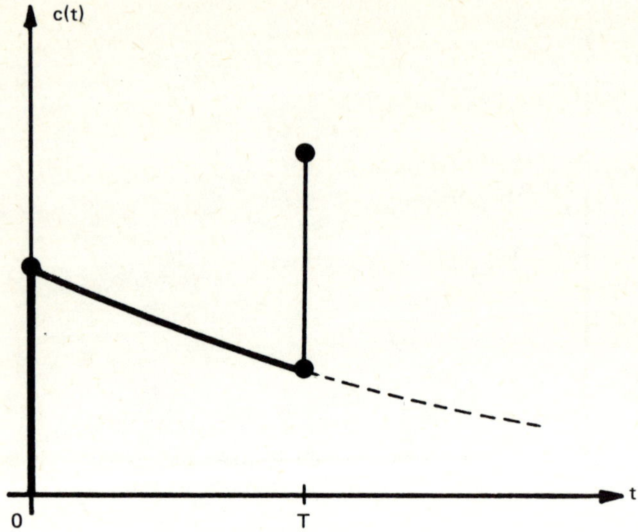

Fig. 18.4

where we assume that the drug is instantaneously dispersed throughout the body. We know that $c(t) = c_0e^{-kt}$, so we have

$$c(T) = c_0\, e^{-kT} \quad \text{and} \quad c(T^+) = c_0 + c_0\, e^{-kT} \quad .$$

This new amount of drug now decays exponentially as before, so that

$$c(t) = c_0(1 + e^{-kT})\, e^{-k(t-T)} \quad \text{for} \quad t > T \; .$$

Thus

$$\begin{aligned} c(2T) &= c_0\,(1 + e^{-kT})\, e^{-kT} \\ &= c_0\,(e^{-kT} + e^{-2kT}) \end{aligned}$$

and $c(2T^+) = c_0\,(1 + e^{-kT} + e^{-2kT})$ after another dose. It is now clear that we can generalise this result to any number of doses. We obtain

$$c(nT^+) = c_0(1 + e^{-kT} + \text{——} + e^{-nkT})$$

which tends to the value $c_0/(1 - e^{-kT})$ for large n.

This has all become rather mathematical, so let's bring it back to earth by trying to decide the dose D and interval T to keep the concentration in the therapeutic range $5 \leqslant c(t) \leqslant 20$. For our 50 kg patient we already have $k = 0.17$/hour, and it seems fairly clear that we want the limit $c_0/(1 - e^{-kT})$ above to be 20 mg/l. Thus $c_0 = 20\,(1 - e^{-kT})$, and so we can choose convenient values for T and obtain the corresponding dose $D = 25\ c_0$. This gives the following table, where we have also included the concentration just before the second dose.

T hours	D mg	$c(T)$ mg/l
1	78	2.6
2	144	4.1
3	200	4.8
4	247	5.0
6	320	4.6
8	372	3.8
12	435	2.3
16	467	1.2
18	477	0.9
24	492	0.3

[Note that we could also have graphed D against T, but in some ways the table above is more useful. This is because the hospital administering the treatment will want a regular timetable for giving the injection, probably every 4, 6, or 8 hours, and so will not be interested in other values.]

We now see that if we choose $T = 4$ hours the concentration *never* leaves the therapeutic range. Thus, unless a concentration just less than 20 mg/l is preferable to one just greater than 5 mg/l, there is no point at all in giving a large initial dose and smaller repeat doses. Our advice to the doctor must be

"Give a dose of 250 mg every 4 hours".

19 RATS

19.1 Introduction

In an experiment by the biologist B. F. Calhoun (reported in *Scientific American*, (1962), **206** 139), a population of wild Norway rats was confined in a ¼-acre enclosure. There was an abundance of food and places to live, and predation and disease were almost eliminated. Thus only the animals' behaviour with respect to one another remained as a factor that might affect the increase in their number.

It was noted that although the adult death rate remained constant at around 5% per month, the number of surviving infants fell as the population density increased. These data were presented as an effective change to the birth rate, so that instead of producing an average 0.4 infants/month each female rat only produced 0.4 K infants/month. The graph of this multiplicative factor K against population density is shown in Fig. 19.1.

Draw a rough sketch of birth rate and death rate (rats/month) against population (rats), and use this to justify the existence of an equilibrium population of about 200 rats. Approximate the graph for K above in a suitable way,

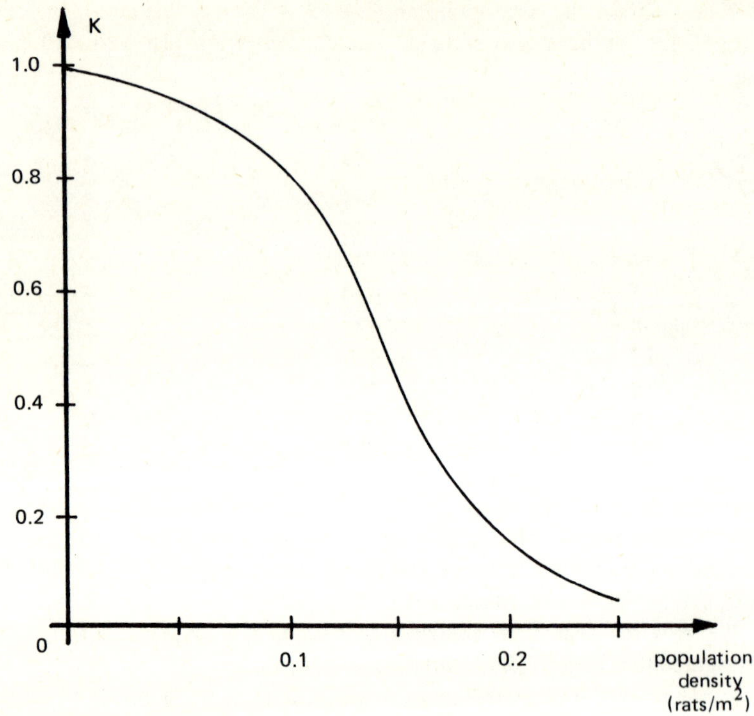

Fig. 19.1

and hence derive a differential equation model. Confirm the size of the equilibrium population, and find out by how much the population has grown in one year.

19.2 Possible solution

In this question the information is presented in months, so we will take this as a natural timescale. We are told that the death rate is 5% per month, so we write

$$DR = 0.05\,P$$

with P being the adult rat population. To write down an expression for the birth rate, we assume – in the absence of any other information – that half the adult population is female. Then the birth rate is given as

$$BR = 0.4\,K\,P/2\ .$$

To be able to sketch BR and DR against the population P we must now look at the graph for K. This is given as K (non-dimensional) against population density (rats/m^2). We also know, however, that the rats are in a ¼-acre enclosure – about 1000 m^2 – so we can change the axis

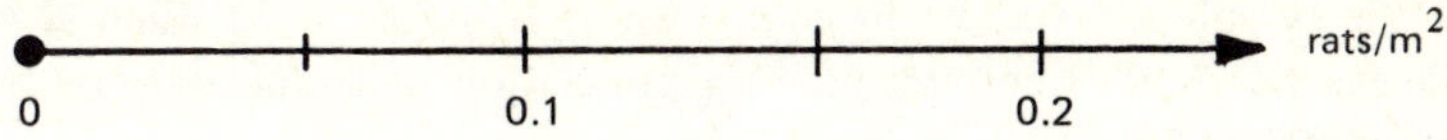

to the axis

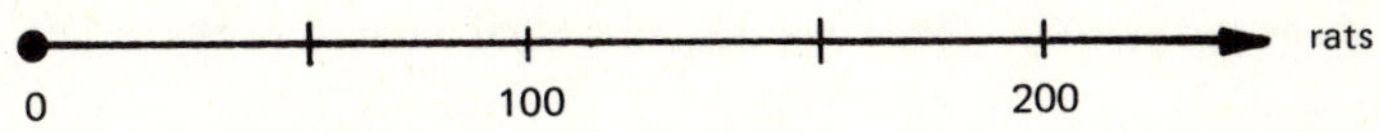

and leave the curve unchanged.

We are now in a position to sketch, very roughly, the birth and death rates against the population (Fig. 19.2). [Here the birthrate curve was obtained by choosing a few values of P, reading off the related value of K, and working out

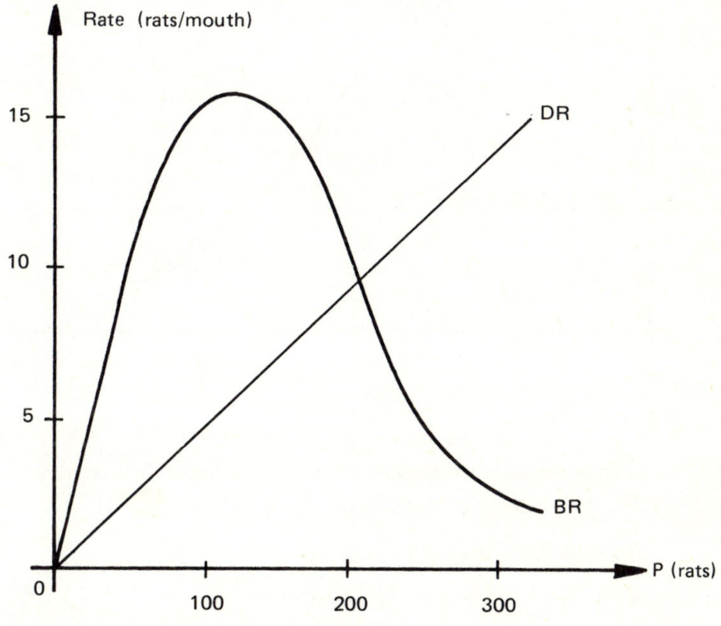

Fig. 19.2

$BR = 0.2\ KP$.] We see at once that $BR = DR$ (and hence the numbers joining and leaving the population balance, leaving the population in equilibrium) when $P \approx 200$. To confirm this result using a differential equation model we can write

$$\begin{aligned} \mathrm{d}P/\mathrm{d}t &= BR - DR \\ &= 0.2\,KP - 0.05\,P \\ &= 0.2\,P(K - 0.25)\ . \end{aligned}$$

Thus the population will be in equilibrium (i.e. $\mathrm{d}P/\mathrm{d}t = 0$) when $K = 0.25$, which can be seen to correspond to a population of about 200 rats from the original data.

The advantage of this latter approach is that we now have an equation which we can solve for the population P at any time t. We must firstly, though, express K in terms of P. Since we are interested mainly in the region around

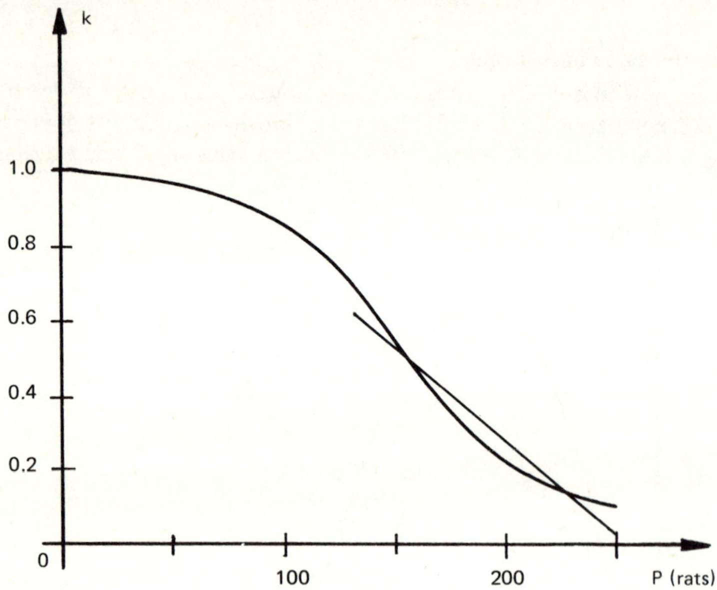

Fig. 19.3

$P = 200$, it seems sensible to approximate K by a linear function *in this region* (Fig. 19.3). We then have something like

$$K = 0.005\,(250 - P)$$

which can be put into our differential equation to give

$$\begin{aligned} \mathrm{d}P/\mathrm{d}t &= 0.2\,P[0.005\,(250 - P) - 0.25] \\ &= 0.001\,P\,(200 - P)\ . \end{aligned}$$

We confirm, again, the equilibrium value of 200 rats.

To answer the last part of the question, and find out the rat population after one year, we need to solve this equation. This is simply enough done, using separation of variables and partial fractions, and we obtain

$$\left| \frac{P}{200 - P} \right| = A\,e^{t/5}\ .$$

Thus $P(t) = \dfrac{200}{1 \pm (A\, e^{t/5})^{-1}}$ for $P \lessgtr 200$, and

writing $P(0) = P_0$ gives $A = \dfrac{P_0}{|200 - P_0|}$.

After one year (i.e. 12 months) the population will be

$$P(12) = \frac{200}{1 \pm (A\, e^{12/5})^{-1}}$$

which is sketched as a multiple of the initial population P_0 in Fig. 19.4. The graph effectively answers the original question.

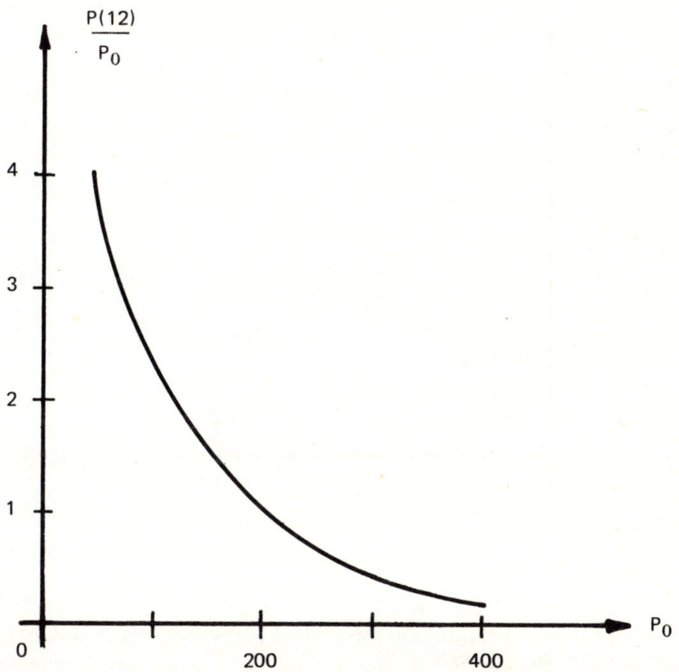

Fig. 19.4

20 A REPLACEMENT POLICY FOR FUEL INJECTION PUMPS

20.1 Introduction

The fuel injection pump on heavy earth-moving equipment has a fairly limited life. If it fails when the equipment is on site the cost of replacement is very much greater – 3 times – than it is if the replacement is made when the machine is in the workshop, since the unit must be transported in for the replacement. It is not possible to examine the pump to determine deterioration likely to cause failure in any very satisfactory way. Hence it is necessary to develop a policy for the replacement time, from new, in order to minimise the costs incurred.

A pump lasts for at most 1 year. Past records show that the probability density function of time to failure has the form shown in Fig. 20.1, so that P [failure in $(t, t + \mathrm{d}t)$] $= f(t)\,\mathrm{d}t$. We wish to choose a time T at which to replace the pump if it has not previously failed. Obviously, the aim is to minimise the cost to the firm. Obtain an optimum value of T analytically. This will involve evaluating the average cost per unit time $c(T)$ and finding the value of T that minimises $c(T)$. Any of a range of methods of varying sophistication and difficulty could be used to do this.

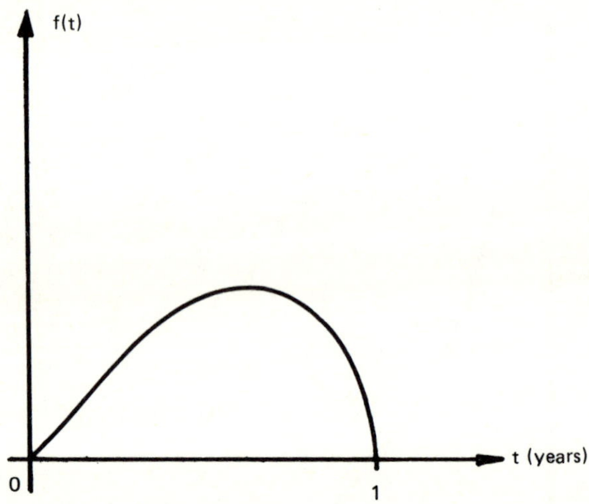

Fig. 20.1

Design and code a simulation model in a high-level language. Obtain as good an estimate of T as possible. Again, any one of a range of methods of varying sophistication might be used, but you will probably need to consider the standard error of your estimate.

20.2 Possible solutions

As is shown in standard texts, the classical analytical approach derives an approximate expression for cost per unit time by finding the average cost of a pump life cycle and dividing it by the average length of a pump life cycle. Pump life cycles are either failure cycles or replacement cycles. The appropriate expression for the average cost per unit time of a replacement policy with renewal at age T is

$$c(T) = \frac{c_F \int_0^T f(t)\,dt + c_P \int_T^\infty f(t)\,dt}{\int_0^T t\,f(t)\,dt + T \int_T^\infty f(t)\,dt},$$

where $c_F = 3 =$ cost of a failure
$c_P = 1 =$ cost of preventative replacement

and for the purposes of this exercise $\int_T^\infty$ is equivalent to $\int_T^1$.

To produce manageable mathematics we approximate the probability density function given in the question by Fig. 20.2, where

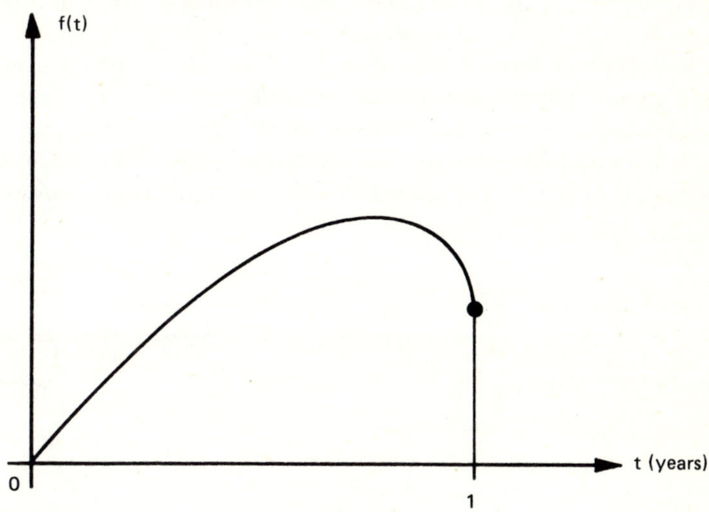

Fig. 20.2

$$f(t) = \frac{1}{5}(18t - 12t^2),\ 0 \leqslant t \leqslant 1,$$

$$= 0 \quad \text{, elsewhere.}$$

This has the correct probability property

$$\int_0^\infty f(t)\,dt = 1 \ ,$$

and yields an average pump lifetime given by $E(t) = \int_0^\infty t\, f(t)\, \mathrm{d}t = 0.6$ years.

It is now a simple job to evaluate the expression for $c(T)$ above. This yields

$$c(T) = \frac{-8T^3 + 18T^2 + 5}{T(T^3 - 3T^2 + 5)} ,$$

which will be minimum when $c'(T) = 0$. This condition turns out to be

$$8T^6 - 36T^5 + 54T^4 - 100T^3 + 135T^2 - 25 = 0 ,$$

which can be solved by the Newton–Raphson method to give

$$T = 0.52 \text{ years and } c(T) = 3.88 .$$

The minimum turns out to be fairly flat, so our advice to the firm would probably be to choose $T = 0.5$ and replace the pumps every six months. The average annual cost would then be minimum, and something around four times the cost of a single preventative replacement.

We were very lucky in this example to be able to approximate $f(t)$ with such a simple expression. In general this would not be the case, and we would use simulation to get an answer. We now repeat the solution this way.

The essential idea of a simulation is that we wish to manufacture a string of fails and preventative replacements with the correct statistics. Thus we may well have

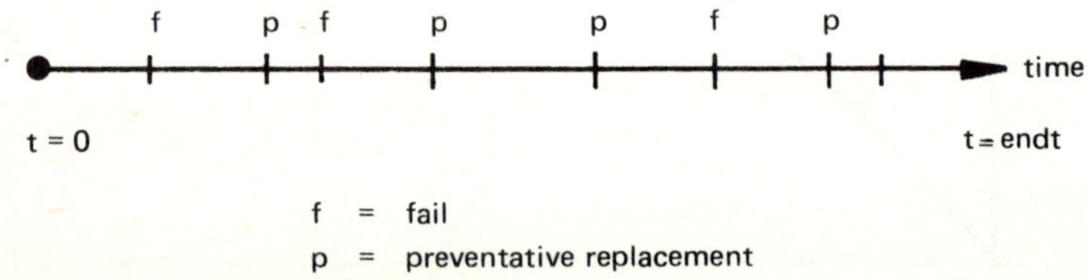

for which we could work out the cost per unit time as

$$\frac{3\text{ (number of fails)} + 1\text{ (number of prevs)}}{\text{endt}} .$$

By repeating this simulation many times (nrepl) we could then work out the average cost per unit time and the associated standard error, and hence have a means of choosing between runs for different values of T.

The procedure outlined above is clearly dependent on being able to generate a typical string of fails and preventative replacements. To do this we must be able to generate a random time (trand) having the probability density function

$f(t)$. This is a standard task in Monte Carlo simulation, and the FORTRAN coding to achieve it is given below.

```
10    trand = 1.0 * rnd (0)
      frand = 1.35 * rnd (0)
      f = 3.6 * trand − 2.4 * trand * trand
      if (trand. gt. f) goto 10
```

[Here the random number generator rnd(0) gives a number with uniform distribution in the range [0, 1] and the expression for f is the approximation used in the first part of the question. The value 1.35 is the maximum value of f.]

The above coding is included in the following program written for a Prime P400, and the sample output that follows uses the values endt = 50 and nrepl = 50. It is clear that we have a very flat minimum, and that the analytic solution $T = 0.52$ has been confirmed. Thus we see the firm has a fair amount of choice in the value of T, and can pick a value to fit best with other maintenance cycles. This may well mean replacing the pump every six months.

```
      WRITE(1,5)
5     FORMAT('INPUT RANDOM SEED, ENDTIME, & NO. OF REPLICATIONS')
      READ(1,*) ISEED,ENDT,NREPL
      WRITE(1,6)
6     FORMAT(////' TIME',9X,'COST ',5X,'S. ERROR')
      WRITE (1,7)
7     FORMAT(' ----',9X,'---- ',5X,'---------'/)
      TSEED=RND(ISEED)
      DO 60 I=1,81,5
      TT=0.09+0.01*I
      TCOST=0
      T2COST=0
      DO 40 J=1,NREPL
      CLOCK=0
      FAIL =0
      PREV=0
10    T=TT
      TRAND=1.0*RND(0)
      FRAND=1.35*RND(0)
      F=3.6*TRAND-2.4*TRAND*TRAND
      IF(FRAND.GT.F) GOTO 10
      STEP=MIN0(T,TRAND)
      CLOCK=CLOCK+STEP
      IF(CLOCK.GE.ENDT) GOTO 30
      TRAND=TRAND-STEP
      T=T-STEP
      IF(TRAND.NE.0) GOTO 20
      FAIL=FAIL+1
      GOTO 10
```

```
20    IF(T.NE.0) GOTO 10
      PREV=PREV+1
      GOTO 10
30    COST=3*FAIL+1*PREV
      TCOST=TCOST+COST
40    T2COST=T2COST+COST*COST
      AV=TCOST/NREPL/ENDT
      SE=SQRT((NREPL*T2COST-TCOST*TCOST)/NREPL/(NREPL-1)/NREPL)/ENDT
      WRITE(1,50) TT,AV,SE
50    FORMAT(F5.2,5X,F8.2,5X,F8.3)
60    CONTINUE
      CALL EXIT
      END
```

```
INPUT RANDOM SEED, ENDTIME, & NO. OF REPLICATIONS
123 50 50
```

TIME	COST	S. ERROR
0.10	10.38	0.012
0.15	7.23	0.018
0.20	5.76	0.023
0.25	4.96	0.027
0.30	4.45	0.028
0.35	4.20	0.037
0.40	3.98	0.035
0.45	3.90	0.046
0.50	3.85	0.043
0.55	3.89	0.045
0.60	3.89	0.041
0.65	3.94	0.045
0.70	4.04	0.045
0.75	4.22	0.042
0.80	4.29	0.041
0.85	4.40	0.046
0.90	4.62	0.041

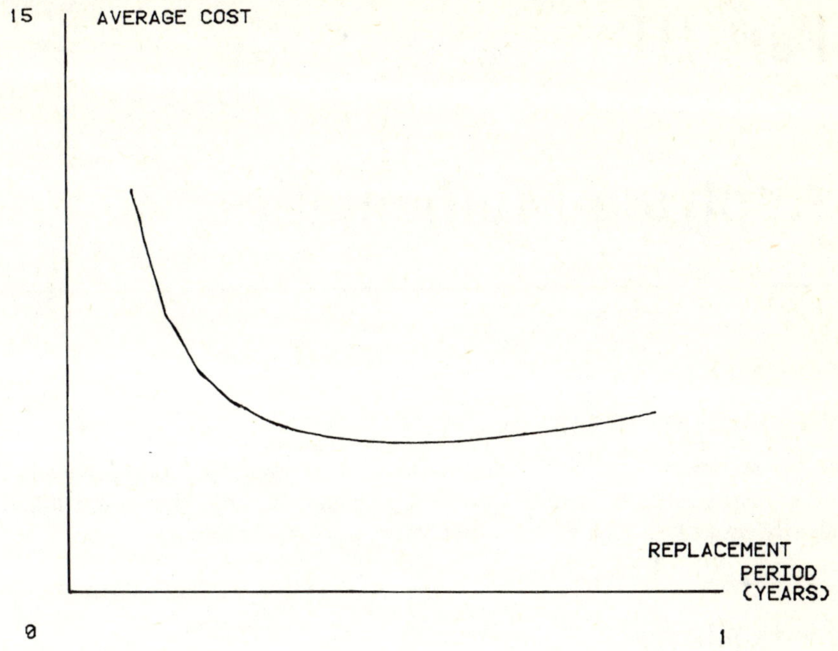
15
AVERAGE COST
REPLACEMENT
PERIOD
(YEARS)
0
1

Part III

Applying Mathematics

INTRODUCTION TO PART III

In this section we attempt to bring out some of the important concepts involved when mathematics is used to solve real problems. We hope that you will have already worked through Parts I and II before reading this section.

1 INTRODUCTION

In Part I of this book we have attempted to show, without doing too much mathematics, that unexpected aspects of our environment, culture, and way of life can often be explained or influenced by mathematics. It is hoped that in working through Part II the reader will have begun to appreciate the diversity of the possible range of applications of mathematics and, by carefully following the case studies, will by now have developed some facility in creating and using mathematical models. In this third Part we will analyse the approaches used in the previous two sections and attempt to give the reader guidelines for applying mathematics. It cannot be emphasised strongly enough that these are guidelines only and that there are no recipes or algorithms for applying mathematics or building models. Successful applied mathematicians, as well as having a sound knowledge of many areas of mathematics, also require to show initiative and flair and, increasingly as mathematics is applied to more varied situations, an ability to communicate with others – mathematicians and non-mathematicians.

Like any other practical skill, skill in applying mathematics is probably acquired best through emulation and practice. We would hope that readers would not attempt to build mathematical models by doggedly following, in sequence, the following sections. What we do hope is that, by reading the following sections, the reader will become more aware of some of the more important steps in the process of applying mathematics – steps which are often unconsciously carried out by experienced applied mathematicians and which are often skipped over in accounts of their work.

2 MATHEMATICS AND THE REAL WORLD

From the examples in Parts I and II it should be clear that, whatever kind of mathematics was used, it was used so that we could develop a clearer understanding of some aspect of what might vaguely be called 'the real world'. The motivation for acquiring this improved understanding can range from pure curiosity – as in the dating of the Table in section 1.3 of Part I to the need to exercise some control over a situation – such as establishing whaling quotas or flight regulations. Whatever the motivation, the real world problem has been transferred from its own environment to an environment in which it may be studied more conveniently. The latter environment, in each case, might be called the mathematical world. Conceptually, we have the picture in Fig. 2.1. Each of these worlds has its own language, and much of the skill and art of the applied

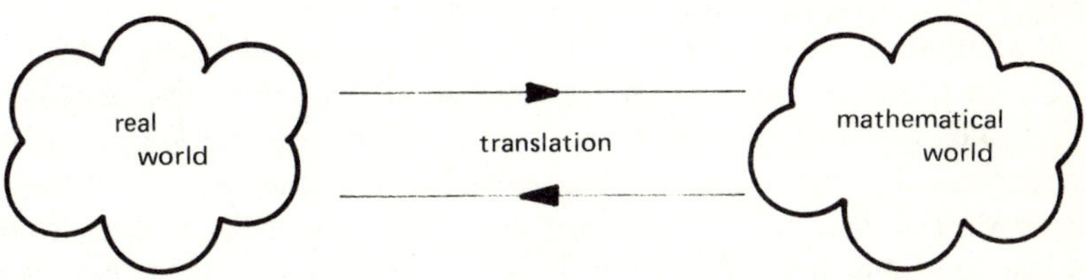

Fig. 2.1

mathematician lies in the ability to carry out the necessary two-way translation process – expressing the real-world problem in mathematical terms and interpreting the results of the mathematical analysis. Although this translation may be difficult, the dividends lie in a key feature of the mathematical world. That feature is abstraction, which might be defined as the art of identifying the common features of different things so that general ideas can be created and applied to different situations. As an example of the potential power of mathematical abstraction, readers should look again at the sections on whaling and King Arthur's table, where they will note that a differential equation of the form $dN/dt = k\,N$ lies at the core of the solutions of the problems. Indeed it is infrequent, and in the context of the problems in this book never, that completely new mathematical ideas and structures have to be created in the solution of practical problems. Having said that, however, we should not lose sight of the fact that many of the major development in mathematics, such as the calculus for instance, were inspired by the need to solve practical problems. Such great leaps, unfortunately, are left to relatively few in the mathematical profession!

3 UNDERSTANDING THE REAL WORLD

An essential step in applying mathematics to the real world is for the mathematician to acquire a thorough understanding of the problem situation. If we stop to think about it, those of us who are numerate – if not mathematically gifted – are continually applying mathematics, albeit of a simple nature, to our daily lives. Since the mathematics we use is at our finger-tips, and we are thoroughly conversant with the situations, we are scarcely aware that we are doing mathematics. Nevertheless, we are using mathematical abstractions to solve our problems. It may be revealing for readers to take some time now to compile a list of occasions, in the last week say, when they have used mathematics (probably arithmetic) in the solution of daily problems, and to analyse the kind of abstraction used and the help it has been. The list will inevitably involve dealings with money – comparing prices, working out discounts, dealing with loans, mortgages and so on. It may involve the kind of planning necessary in 'do it yourself' jobs – measuring, converting, pricing, ordering. It will also, inevitably, include situation in which the reader has to cope with time – making transport connections, for instance.

As an illustration consider the following situation. A rail traveller wishes to travel from a Glasgow suburban station (Hamilton) in the morning, arriving at Leeds before 12.00 hrs the same day. Consulting the information office, the traveller finds that the only direct train serving this purpose is the 07.05 from Glasgow via Dumfries, see Fig. 2.2, arriving at Leeds at 11.52. A check of the suburban timetable reveals that the 06.30 from Hamilton arrives Glasgow at 06.35 – just about allowing time for the connection (allowing for some delay and change of platforms). This is not particularly good news since it means a 05.45 rise. The passenger then realises that the train to Leeds follows a rather circuitous route to Carlisle, which it leaves at 09.30, and that there is a train leaving Glasgow for Birmingham at 07.40. A better route, therefore, might be to go from Hamilton to Motherwell for the Birmingham train and change at Carlisle for the 09.30 to Leeds! This means a 06.40 rise, which is a bit more civilised.

Many readers may not consider this to be a problem at all, far less a mathematical problem. It should not, however, be dismissed out of hand, because there are many intelligent but non-numerate people who cannot cope with the mass of data and calculations involved in exercises such as the above. Looking at the example again, we see that it involves the following stages, some of which are not at all simple.

(i) A thorough local understanding of geography and rail networks.
(ii) The ability to collate information from three railway timetables.
(iii) The estimation of platform transfer times.
(iv) Allowances for error.
(v) Some simple calculations.

The main lesson to be derived from this discussion is that, in applying mathematics to a practical problem, it is important to acquire a knowledge of the problem situation akin to the local knowledge required for the travel problem. In that problem the mathematical techniques used were simple, and the success

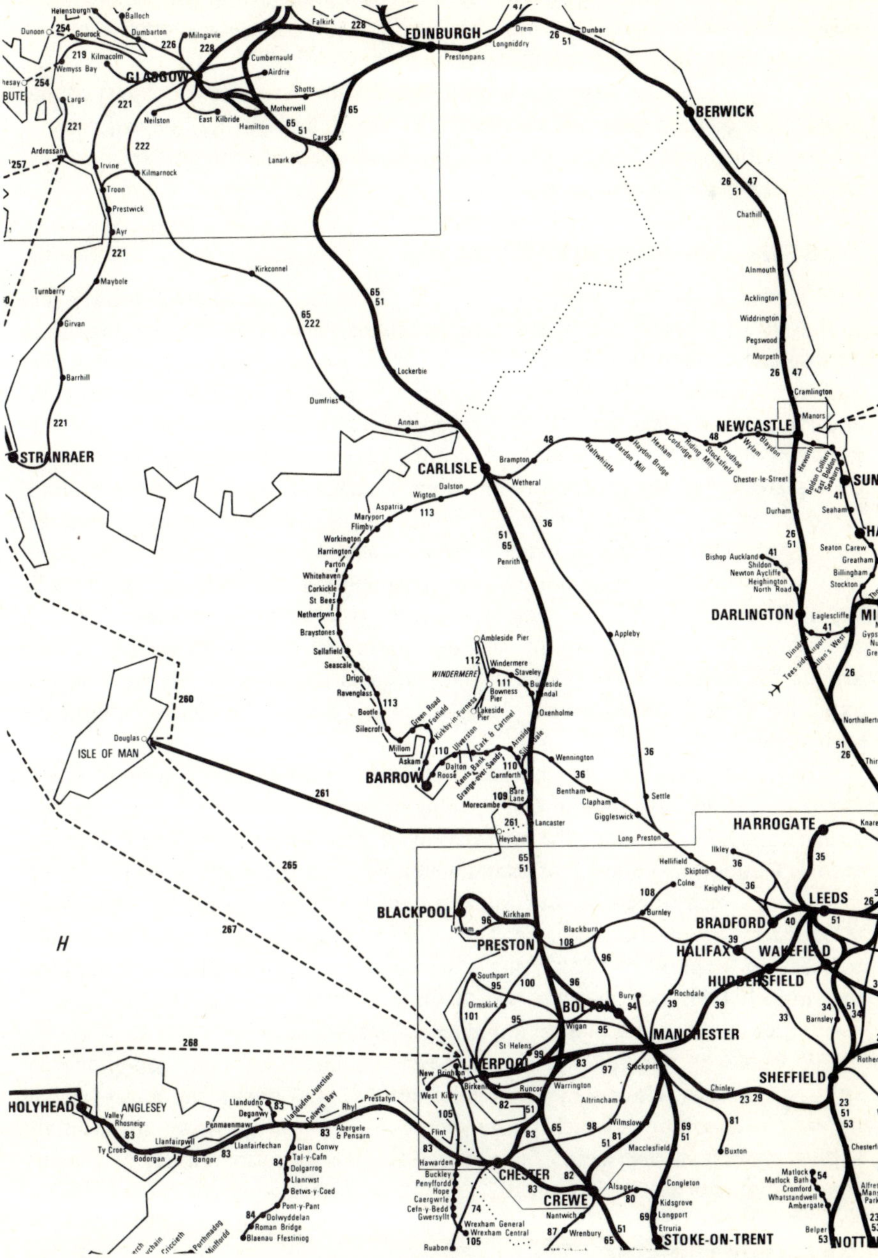

Reprinted with the permission of the British Railways Board

Fig. 2.2

of the exercise lay in the understanding of the problem. At this stage readers may find it discouraging to realise that a skilled and experienced applied mathematician might find most of the problems in this book as transparent as the reader might find the passenger's problem. On the other hand, readers should now find encouragement and motivation in the realisation that the difficulty in dealing with practical problems often lies in the understanding of the problem rather than the mathematics used.

4 GETTING TO GRIPS WITH PROBLEMS

One notion that today's aspiring applied mathematician should quickly lose is that it is a career which can successfully be enjoyed in the contemplative atmosphere of an office. Recall Fig. 2.1, where we saw that the mathematician has to be involved in a two way process of translation between the real world and the mathematical world. This process inevitably involves the mathematician in *communication* with others. Real problems are rarely given to mathematicians on a plate. Often the professional mathematician will be acting as a consultant – either in an 'in house' capacity within an organisation, or as an external consultant. The problem is often presented verbally and, even when the statement of the problem is supported by written material, the initial picture may be vague. There may be conflicting interests at play, as in the design of aircraft routes or the establishment of whaling quotas in Part I. As well as reading around the subject of the problem, the mathematician has to talk with people, exploring their knowledge and ideas, collecting facts, following up alternative approaches.

As an instance of this, consider the example on aircraft collisions. It was almost ten years between the first brewing of trouble between the pilots and the airlines and the eventual implementation of the conclusions drawn from the model. During this period Professor Machol, who was a consultant to ICAO, was involved in many hearings and international meetings – some of these being attended by about 100 delegates. The technical problems were thrashed out and solved over a period of three years – which included six of these international meetings. It then took another two to three years to 'sell' the solution to all of the parties involved and then to implement the system.

It would be wrong to imply that every applied mathematician sooner or later gets to deal with problems of international importance and always comes up with neat solutions which please everyone. However, in many instances the applied mathematician has to be prepared to question, listen, argue and, sometimes, eventually act as a salesman.

5 SYSTEMS THINKING

Realising that problems are frequently only vaguely presented and that skills in communication are important to the applied mathematician, it is reasonable to ask whether there are guidelines for practical problem solving. This is an area in which mathematicians will differ in opinion. Some will argue that the process of practical problem solving can be formalised, and others will argue that it is an intuitive art. One thing is certain, and that is that the task cannot be approached in a haphazard way. It is important to solve the right problem, and this is not so easy as it may seem. If we are only handling one specific part of a general problem area it is important that we realise that. For instance, in solving the airline routing problem it was important that the consultant for the pilots also took into account the difficulties of the airline operators – otherwise he may well have produced a solution acceptable to the pilots but which would have been dismissed out of hand by the airline companies. Again, if economic considerations and pressures are required in developing conservation models for whales, then it is important that this is recognised by the model builder.

One approach to problem solving which attempts to identify the problem clearly within its context is something called 'systems thinking'. At its simplest level – which is all we will be concerned with in this book – systems thinking is just applied commonsense, although there are one or two concepts which are worth formalising here.

One such concept is that of a system itself. The word system is one which is convenient but overused in today's society. The *Concise Oxford Dictionary's* definition is "a complex whole, set of connected things or parts, organised body of material or immaterial things . . .", and we frequently use the word this way in everyday life – digestive system, solar system, nervous system, railway system, computer system and so on. The term thus covers a broad spectrum of our physical, biological, and social world, and also conveys the idea of complexity.

In applying mathematics to a problem situation, it is often convenient to call the situation we are studying a system. The section on the attractiveness of competing shopping centres, for instance, is a good example of a situation where we are dealing with a system consisting of many components. Most systems contain subsystems and also themselves are subsystems of larger systems. Within a particular system, the subsystems will react with one another. Although this seems abstract, it is really important to appreciate this in solving problems. For various reasons we often have to limit the scope of our study, and in such cases we have to know where to set up the boundary lines. An example of a real problem will help to illustrate these ideas.

6 SYSTEMS THINKING – AN EXAMPLE

Consider the following scenario (based on an actual situation). A fairly large local authority has, as one of its functions, the responsibility of providing an efficient waste management system for both domestic and industrial/commercial premises. Thus it must provide facilities for collecting and disposing of various types of waste originating from locations spread over a wide area. The collection process involves a large number of technically complex vehicles and a large workforce. The disposal system also includes a variety of facilities, such as incinerators and landfill sites. The management problem includes the creation of collection routes, the assignment of waste to disposal facilities, the deployment of workmen, strategies relating to the purchase, replacement, and maintenance of vehicles, and so on. It was decided to look closely at the nature and level of the existing provision of facilities, and the Management Services Department has approached a group of consultants and has given them the following remit.

> "Carry out an investigation of the waste management system and report on the feasibility of using mathematical models, if necessary computer-based, to help the authority to
>
> (a) improve the efficiency and effectiveness of the waste management system in the short term, and
> (b) react to changing environmental, technical, and social factors and thus maintain an efficient and effective provision in the long term."

As this particular investigation proceeded, the consultants quickly became aware that the total system was quite complex. Even in terms of organisation, there were many overlaps. The Management Services Department was responsible for the current investigation, the Cleansing Department was in the front line of providing the service, the Transport Department was responsible for the maintenance of vehicles, and the Finance Department was responsible for budgetary control. It was difficult therefore for any one member of management in the Authority to have a clear picture of the system and the possible consequences of any one decision. In fact, it was one of the consultants' conclusions that "A model-based approach should, in this context, be directed primarily towards helping management to make decision more quickly and confidently at the tactical and strategic levels. That is, the immediate objective of the model would not be cost minimisation in any particular sector, rather cost reduction should follow from improved flows of information and consequent better control".

In discussing the outcome of the study with the local authority management team it was useful for the consultants to present, visually, their picture of the system and the possible model structure which might subsequently be used. A simplified systems picture is shown in Fig. 6.1.

Figure 6.1 simply emphasises that the local authority is a complex system reacting with its environment – influencing it by its policies and, in turn, being influenced by social and technological changes in the environment. Within the local authority system there are many subsystems, but the ones on which attention was focussed in this study were the collection system (involving the generators of waste, the personnel involved in collection, and collection vehicles), the disposal system (involving disposal plants, landfill sites, personnel, etc.), the

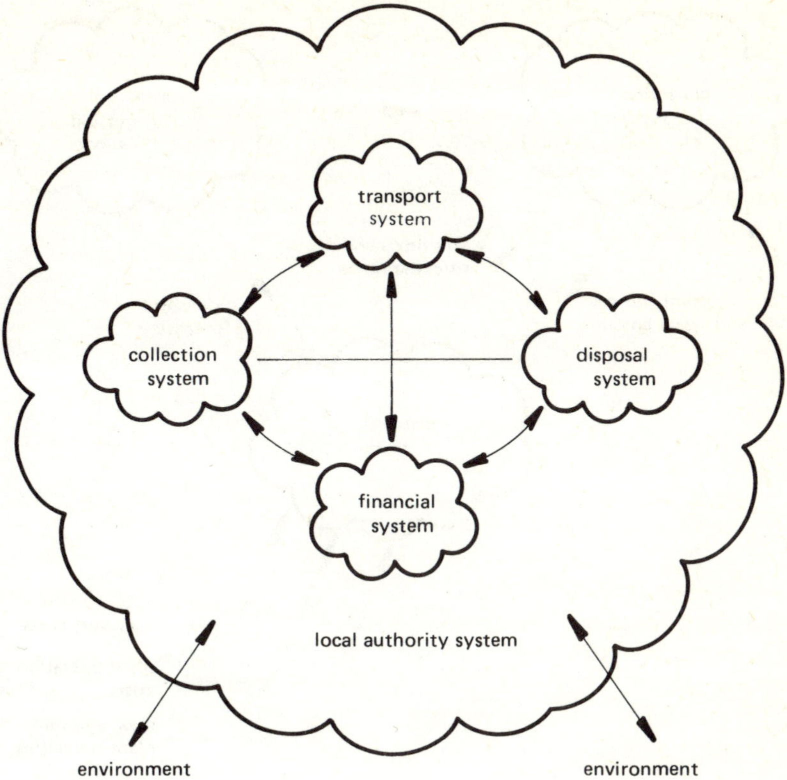

Fig. 6.1

financial system, and the transport system. Each of these systems is influenced by each of the others, and this is emphasised by the two-way arrows linking the system blocks.

Admittedly none of this looks a lot, but it does help the mathematician, when he comes to think about the contribution of models, to place the model idea in context. In this way he also begins to place demarcation lines on the scope of the model and gets the first hints as to what factors should be included. The next stage, therefore, is summarised in Fig. 6.2. In this, the consultants described their idea that there should be three linked models.

The purpose of each model may be summarised as follows.

Disposal Sector: Overall objective of disposal sector would be to identify optimal waste flows and corresponding costs from generation sites to disposal plants (actual and hypothetical). The model would help management to devise strategies for the following situations:

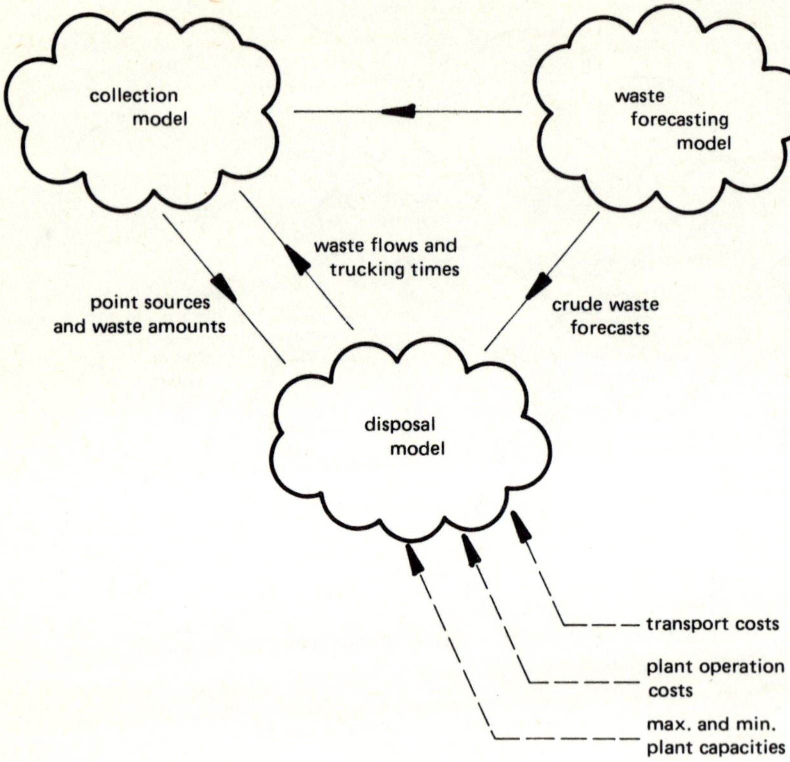

Fig. 6.2

- changes in the amount of waste generated
- planned or unplanned temporary or permanent shutdown of a plant
- the offer of an option on a new landfill site
- the possibility of plant replacement or extension
- the possibility of sharing facilities with another authority.

Collection Sector: Aim of this sector model (or models) would be to allow the rapid evaluation by management of alternative strategies relating to the following:

- manpower availabilities
- vehicle numbers
- vehicle types and capacities
- vehicle maintenance, refurbishment and replacement schedules
- collection team sizes
- sectoring of collection areas
- route construction.

Waste Forecasting Sector: Objective of this model would be to indicate changes in volume and type of waste based on past trends, demographic changes, changes in packing technology etc.

This example has been fairly lengthy, but it illustrates a number of important general points about developing models.

(1) It is possible to move from a very vague description of the problem situation to a fairly specific description of the rôle of the model without introducing any mathematical language. In fact it is usually wrong to decide on the mathematical structure of the model and then attempt to force this structure on the problem.
(2) Having reached the stage outlined in (1), the applied mathematician will have a good idea of most of the factors to be included in the model, and perhaps some of the relationships between some of the factors. The data requirements will also be fairly evident by this stage.
(3) Where the situation is complex and the model has a number of well-defined parts, this approach, by examining the overall structure of the model and the relationship between the parts, allows a modular approach to the model construction. In other words, each part can be tackled one at a time by a single mathematician or each by a different person. Either way, it should be possible to integrate the different parts.

Summary so far

The story so far may be summarised in the form of some specific pieces of advice.

(i) Take time to get to know the problem situation really well.
(ii) If appropriate, get out of the office and have a look at the problem situation. Speak to the people involved.
(iii) Together with your 'client', reach a clear, agreed statement of the problem.
(iv) Decide on the aspects of the problem with which your model is going to deal.

The above process is sometimes called problem *formulation*. From this point on it is possible to start piecing the model together.

7 CONSTRUCTING MODELS

By this stage readers will have seen a number of mathematical models and will recognise that a model is a set of statements about a set of variables. These statements are normally in the form of mathematical relationships between the variables, and by analysing these relationships we are able to make further statements about the variables in the model.

For example, in our waste management situation, a possible model for the disposal sector might be the following. In this model the collection areas have been reduced to point sources of waste – denoted by subscript i – and it is assumed that this waste can be transported directly to landfill sites – denoted by subscript j – or taken to processing plants – denoted by subscript k. Residual waste from the processing plants is disposed of in landfill sites. Figure 7.1 illustrates this situation.

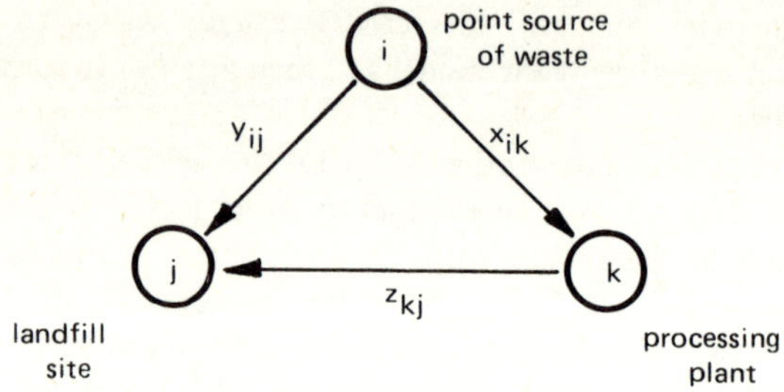

Fig. 7.1

In the model and Fig. 7.1, the variables have the following significance.

i = Point source of solid waste generation S_i $(i = 1 \ldots m)$
k = Solid waste processing plant $(k = 1 \ldots r)$
j = Sanitary landfill $(j = 1 \ldots n)$
x_{ik}, y_{ij}, z_{kj} = Flow of solid waste (tonne/unit time)
$d_{ij}^{1}, d_{ik}^{2}, d_{kj}^{3}$ = Unit transportation costs (£/tonne)
c_j^{1}, c_k^{2} = Unit operation costs (£/tonne)
$100\, r_k$ = Weight reduction of process k

Finally, the model consists of an expression for the cost of operation of the disposal system – which has to be minimised – and a set of relationships which represents the physical constraints and material balances in the system.

Minimise:

$$\sum_{i=1}^{m} \sum_{j=1}^{n} d_{ij}^{1}\, y_{ij} + \sum_{i=1}^{m} \sum_{k=1}^{r} d_{ik}^{2}\, x_{ik} + \sum_{k=1}^{r} \sum_{j=1}^{n} d_{kj}^{3}\, z_{kj} \qquad \text{transport costs}$$

$$+ \underbrace{\sum_{k=1}^{r} c_k^{2} \left[\sum_{i=1}^{m} x_{ik} \right]}_{\text{process}} + \underbrace{\sum_{j=1}^{n} c_j^{1} \left[\sum_{k=1}^{r} z_{kj} + \sum_{i=1}^{m} y_{ij} \right]}_{\text{landfill}} \qquad \text{operation costs}$$

subject to:

(1) $\sum_{j=1}^{n} y_{ij} + \sum_{k=1}^{r} x_{ik} = S_i$ mass balance of S_i in area i

(2) $(1 - r_k) \sum_{i=1}^{m} x_{ij} - \sum_{j=1}^{n} z_{kj} = 0$ modified Kirchoff Law for process k

(3) $\sum_{i=1}^{m} x_{ik} \leqslant CP_k$ maximum capacity of process k

(4) $\sum_{i=1}^{m} x_{ik} \geqslant MCP_k$ minimum load for process k

(5) $\sum_{i=1}^{m} y_{ij} + \sum_{k=1}^{r} z_{kj} \leqslant L_j$ maximum capacity of landfill j

(6) $x_{ik}, y_{ij}, z_{kj} \geqslant 0$ non-negative variables

The model expresses the cost of operation of the waste disposal sector in terms of several key decision (or control) variables. The model also expresses, in mathematical language, the natural constraints which govern the operation of the system. Constraint equation (1), for example, says that the total amount of waste, S_i, produced in area i has to be disposed of either directly, in landfill sites, or via processing plants. Constraint equation (2) simply says that everything which goes into a processing plant must come out again. The latter condition is similar to Kirchoff's Law for an electrical network. The other constraints have similar interpretations.

To the mathematician, the model has a recognisable form – it is an integer programming problem of a particular type. This particular type of mathematical problem can be analysed using standard computer packages. The outcome of the analysis will be information on the present and future status of the system. For example, an analysis of the model might show that the existing system can cope with the amount of waste being generated, but that it could be operated at lower cost by eliminating one disposal plant and putting two others on to three-shift working. It might also indicate that, because of forecast increases in the volume of waste being produced, the system would cease to be able to cope within eight years. Of a number of options available, the cheapest may involve the acquisition of a specific landfill site on which the local authority currently holds the option.

The model would also provide all the necessary information on transport requirements and redistribution of waste to disposal plants and sites.

It would seem, therefore, that there are two things which have to be done in piecing together the model. These are the following.

A The identification of the variables which will appear in the model.
B The establishment of relationships between the variables.

We now discuss each of these in turn.

A Identifying the variables

In reading accounts or descriptions of mathematical models, it often appears as though the writer has 'drawn the variables out of the air'. In some cases this is almost what has happened – since the model-builder is so familiar with the problem situation, the important variables occur quite naturally. Even in these kinds of situations the model builder probably does, at least subconsciously, follow one of two processes. In the first of these, having focused attention on a certain variable, we would attempt to identify what other variables can influence it. In the other process, we would again direct attention on a particular variable, but would then look for other variables which were influenced by that variable. In many problems we would use both processes, thus building up a list of variables. Simultaneously, of course, we would also develop some idea of the relationships between the variables.

Examples

(1) In analysing the motion of the shot putter, it is immediately clear that it is the *range* of the shot which is ultimately important. Thereafter it is fairly obvious that the range is influenced by the *initial velocity of projection* and by the *angle of projection*. It is also clear that *gravity* will have an effect. It might require a little more thought to identify the *height* of the point of initial projection as a relevant factor.

These variables are the only ones which appear in the two models developed earlier. Somewhat deeper thought would be required to identify the factors which contribute to the initial velocity, and here it is often a matter of fine judgement to decide where to draw the line in thinking of more variables which might be relevant. A few model builders believe in throwing up as many variables as possible at the start of the investigation, but probably most believe in keeping the model simple and only introducing new factors as a need for these becomes apparent. The latter approach has been used in the shot putt model with some success because, of the initial variables (velocity and angle of projection) considered, the model has identified velocity as being worthy of greater attention.

(2) In some situations there will be a chain of variables which we must chase through. For example, in our investigation of pricing and advertising policies,

we start off by asking why *price* and *advertising expenditure* are important. They are important because they influence *profit*, but what is the exact mechanism?

To obtain an understanding of the mechanism we can turn to the well-known expression for *profit* – *income* less *expenditure* – and then ask how price and advertising expenditure influence each of these. Following this line of reasoning, we can end up with a network of variables (Fig. 7.2), without as yet establishing the exact relationships between the variables.

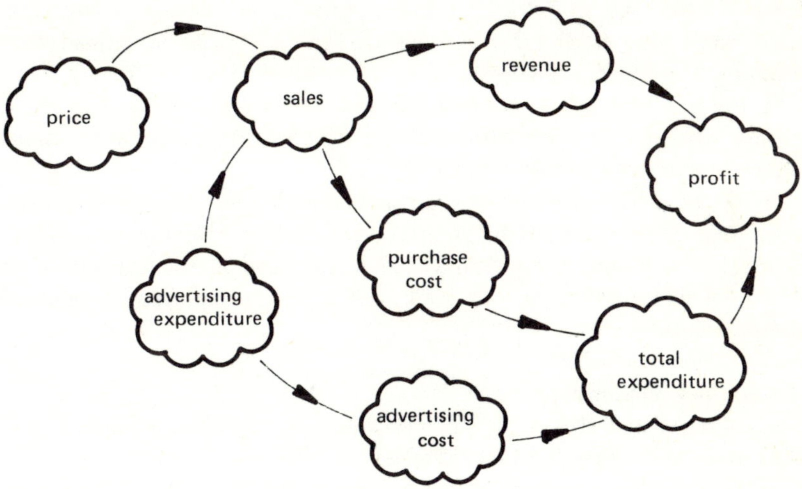

Fig. 7.2

(3) In other cases, we may, right from the start, identify several variables as key variables. Yet some of these may disappear as we proceed with the model building process because of the assumptions we make in developing the model. A good example of this is the stock-holding problem where, at the start, *size of orders* placed, *interval of time* between the placing of orders, *demand* for the product, are all, clearly, determinants of the cost of operating the system. There are thus, potentially, many variables in the model.

Assumptions made in developing the model – and the control system – limit the number of variables in the model. The assumption of steady demand and the decision to operate a replenishment system with a constant batch quantity reduce the number of variables to a single variable – the replenishment quantity.

So far we have been looking at the conceptual identification of variables, that is the realisation that a particular factor is relevant and can ultimately be represented in the model by a variable. Before leaving variable identification, however, it is worth pointing out that the conceptual identification is not always the end of the story. To be useful in a model, a variable has to be capable of clear definition and measurement – and this is not always as easy as it seems.

In Part I we discussed the role of mathematical models in the formulation of international whaling policy. One of the problems inherent in making policies lies in measuring whale populations. Conceptually, therefore, populations level is seen as an important variable. It is not easy, however, to measure whale populations, so one of the methods used is to attempt to relate population to catch using the concept of 'catch/unit effort'. This gets over one difficulty of measurement but raises others. Effort has to be expressed in terms of a unit which has the same value at all times and places. The unit might be defined as the operation of a vessel, gunner, and crew of standard efficiency in a unit area of sea for unit time. Units of time and area are easy enough to handle, but 'standard efficiency' is more difficult to quantify because it depends on variable characteristics of vessel, crew and gunner.

We do not wish to labour the point, and hope that this example is sufficient to emphasis the distinction between conceptual and operational definitions of variables. It is not a problem which looms large in the context of the relatively simple problems in this book, but it is very important in many real world problems.

B Establishing relationships

In the course of identifying the variables which are likely to be relevant to the model, it is likely that the model-builder will also develop some idea of the relationships between the variables. At that or some subsequent stage, the relationship will have to be developed more formally.

In some cases the mathematical relationship will be very obvious, for example

$$P = R - C$$

where P is profit, R is revenue and C is cost.

Sometimes, particularly in the modelling of physical systems, mathematical relationships in the model will follow from well-known laws such as Newton's Laws of Motion. In our analysis of shot putt performance (section 14 in Part II), for example, once a few assumptions have been made concerning the extent of the model, the important relationships in the model follow from the application of laws of motion.

In many cases the model-builder may be convinced that there is a relationship between some of the variables in the model, but may be quite some way from finalising the mathematical relationship. This was the case when we looked at the attractiveness of competing shopping centres earlier. At one stage in the creation of the model we introduced the probability, T, of an individual making a trip to a particular shopping centre. It became apparent that T would depend on the distance, D, of the individual from the centre and the attractiveness, F, of the centre. Mathematically speaking, at this stage all we have is

$$T = f(D,F)$$

where f is some function still to be identified.

So, how do we decide on the form of f? Again, as we have mentioned before, there is some merit in looking for a simple relationship. A little thought will lead us to the conclusions that T will increase as F increases and will decrease as D increases. We are therefore a little further forward, in that we can say that T varies directly with F and inversely with D. However, powers of F and D could equally well be involved in the expression for f. Ultimately we used the analogy of the attraction between two particles of matter which, according to Newton's Law of Gravitation, varies directly as the product of the masses of the particles and inversely as the *square* of the distance between them. We thus take f to be directly proportional to F (because this is the simplest relationship and there is no evidence to encourage us to use anything more complex) and inversely proportional to D^2 (because of the analogy outlined above). We thus ended up with

$$T = \frac{KF}{D^2} ,$$

where K is the constant of proportionality. This kind of relationship might be called *explanatory* because it attempts to give a mathematical explanation of the relationship between the variables, based on some understanding of the underlying mechanism.

Sometimes it is not possible to provide an explanatory relationship, or it might not be worth the effort or cost of developing a clear explanatory relationship in a particular situation. In our study of the retail price index (section 7 in Part II), we used a moving average of previous index values to provide forecasts of forthcoming new values. Whatever formula of this type is used, it makes no attempt to *explain* changes in the index in terms of changes in factors which give rise to variations in the index. It is probably satisfactory provided there are no really radical changes in the factors which influence the index. As such, it would probably be useful to a trade union using forecasts of the index as part of its case for an increase in wages. It would be completely useless to Treasury officials who might want to forecast the effect of certain government or international decisions on the index. The latter kind of relationship is said to be *empirical.*

One of the great dangers in using a mathematical model is when we forget its limitations. Thus if our model is based on empirical relationships and we try to use it for predictive purposes, there is the danger that the underlying factors may suddenly change. It is therefore difficult to have complete confidence in empirical models in certain circumstances. A classical example of this is described by Blackett in his book *Studies of War.* During the Second World War convoys of merchant ships were escorted between Britain and the USA and Russia by warships. There was a limited number of escort ships available for convoy duty, and some thought was given to the optimum configuration of convoys. The prevailing attitude was that smaller convoys were relatively safer than larger convoys. A convoy of 40 merchant vessels was considered to be optimum, and convoys of over 60 ships were prohibited. The escort size was given roughly by the rule

$$\text{escort size} = 3 + \frac{N}{10},$$

where N was the number of merchant ships. This rule – of unknown origin – was in itself odd, since three convoys of 20 ships would require a total of 15 escort ships whereas one convoy of 60 ships would be allocated only 9 escort ships.

As part of this investigation mathematical models were used in an attempt to support or refute the contention that large convoys were safer than small ones. A statistical analysis of two years' data indicated that large convoys were more than twice as safe as small ones, but the model-builders were hesitant to suggest a change in policy from the existing small convoys to large ones. Their hesitancy was due, in part, to the possibility of large statistical errors, and, in part, to the possible consequences in terms of human life of an error arising from the model. They therefore sought an *explanation* of the statistical results. The following statements seemed to provide the basis of an explanatory model.

(i) The probability of a U-boat seeing a convoy is independent of convoy size.
(ii) The probability of a U-boat penetrating the escort is dependent on the escort density around the perimeter of the convoy.
(iii) The expected number of ships in the convoy sunk is independent of the convoy size once the escort is penetrated.

Hence for a fixed density around the perimeter, the expected number of ships sunk is independent of the convoy size. Hence the proportion of ships sunk is inversely proportional to the convoy size.

With the support of this analysis, the decision was taken 'after some weeks of earnest argument' to alter the regulations on convoy size. This decision was taken in the spring of 1943, and from then on the average size of convoys grew. Blackett, quite rightly, makes no claims that the analysis was vindicated by the reduced losses because, as he says, so many other factors were changing. He does note, however, that one convoy of 187 ships arrived safely in the summer of 1944. He also emphasises the fact that it was as a consequence of this analysis that anti-submarine escort vessels were diverted from convoy duty to support the invasion of Normandy in 1944. Assuming that it was the change of strategy which led to the reduction in losses, 200 ships could have been saved had the strategy been adopted even a year earlier.

8 CHOICE OF MODEL FORM

In the previous section, we examined the variety of ways in which relationships can be created for use in mathematical models. It is appropriate at this stage also to note that, in some situations, there may well be some flexibility in the choice of the form of the model itself. There may well be the choice as to whether the model should take one of the following forms:

Deterministic
Stochastic
Simulation
Other form?

9 MODEL VALIDATION

At various stages in the process of constructing and using a mathematical model it is important to check that the model behaves as expected and that it reflects adequately the real system which is being modelled. This testing of the model is called *validation,* and it can be carried out at the following stages.

During model construction
On completion of the model
On implementation of the model.

The ultimate validation of a model is clearly carried out at the implementation stage, when we ask if the model fulfils the purpose for which it was created. Equally well, there are certain dangers inherent in leaving the validation of the model to this late stage.

Validation by implementation can be a traumatic experience for the mathematician and his client. There are, however, many circumstances when this is the only real form of testing which is available. This is particularly true when models are being applied in quite new situations. In a previous section we discussed the use of mathematical models in formulating policy on large convoys. Ultimately, the only way of testing the prescription of this model was to form large convoys and observe the results – a nerve-racking experience for those involved in making the decision. If more data were available, or if one had the facility for building some physical models, one would be able to test a model on completion but prior to implementation. There is a whole variety of ways in which this kind of testing can be done, but we will confine ourselves to only two examples.

In our modelling of the movements in the retail price index, it was possible to check our models against the extensive data available. We did this by testing each model's ability to 'predict' known index values. If the models cannot produce tolerable estimates of known index values, then there is little point in using them to forecast future movement of the index. Of course, as we commented earlier, the fact that the models do produce good estimates of past index values does not guarantee their reliability in forward forcasting.

In Part II we also looked at some elementary stock control, including the well-known EBQ formula

$$Q = \sqrt{\frac{2 N c_1}{a c_2}}$$

where Q is the Economic Batch Quantity
N is annual demand
c_1 is the order cost
and ac_2 is the stockholding cost.

In an actual situation, a light-engineering company invited consultants to advise them on the introduction of a computerised stock control system. The company had been advised by its accountants that it was carrying too much stock and that the auditors were likely to devalue old stock. Hitherto the company had operated a fairly *ad hoc* stock replenishment policy and the consultants, after spending some time studying the system, decided that the simple EBQ formula would provide a basis for a stock replenishment policy. There was a large amount of data, going back many years, on customer demand, orders for stock replenishment and stock levels. The consultants decided to test the EBQ formula by comparing the use of the EBQ-based policy with the existing policy, using two years' historical data on a number of stock items.

Demand figures were available from old stock record cards, so all the consultants needed for a calculation of the EBQ was the order cost (c_1) and the stockholding cost (ac_2). The company did not seem to have these figures available but did some quick calculations to provide values of c_1 and ac_2. The consultants then used these figures in the EBQ formula, and the first thing they noticed was that the EBQ was much higher than any of the orders which had been placed for replenishment stock in the two-year period used in the test. A policy based on the use of EBQ was tested using the two-year historical demand data, and the cost of this policy was less than that of the existing policy. The worrying thing was that the new policy involved much *higher* stock levels rather than *lower* levels which the accountants wanted. It was easy to see why this had happened. It was simply that the order cost was high in relation to the holding cost. In view of the company's initial uncertainty about order and holding costs, the consultants went back and asked them to think again about these figures and the consequence which was higher stock levels. The company replied a day or so later, with some embarrassment, that it had miscalculated the order cost – it should have been lower by a factor of about 10!

The point, in this case, was that it was not the model which was wrong but the data used with the model. The lesson is that it is not just the model, but the model in relation to the system, which needs to be validated in some cases before putting the model into action.

Many of the difficulties which might arise in the validation of models at the completion or implementation stage can be avoided by painstaking testing of the model during the construction stage. This includes testing of individual relationships and testing of data. In Part I, for example, we showed how the differential equation model

$$\mathrm{d}N/\mathrm{d}t \;=\; [r-(M+F)]\,N$$

could be used to investigate the effect of fishing (represented by the parameter F) on the size, N, of a population of whales. The reliability of the model depends, amongst other things, on the reliability of the values used for the parameters r and M. In our example we used fixed values for r and M, but in the study on which the study is based several different methods were used to provide a range of possible values for r and M.

10 SUMMARY

The process of applying mathematics to real world problems may be summarised in flowchart form (Fig. 10.1). In the end, the implementation of the model will tell us more about the real system or, perhaps, the model will be used directly in the form of some kind of automatic control of the system. Whatever the purpose of the model, a kind of feedback process will have been completed.

Again we must emphasis that Fig. 10.1 is provided to give the reader an overall impression of the whole process of applying mathematics to real problems via the mechanism of mathematical models. Figure 10.1 shows key steps in this process, but the impression should not be left that these steps are carried out in a neatly ordered way. Even for the best organised applied mathematician there will be much retracing of steps.

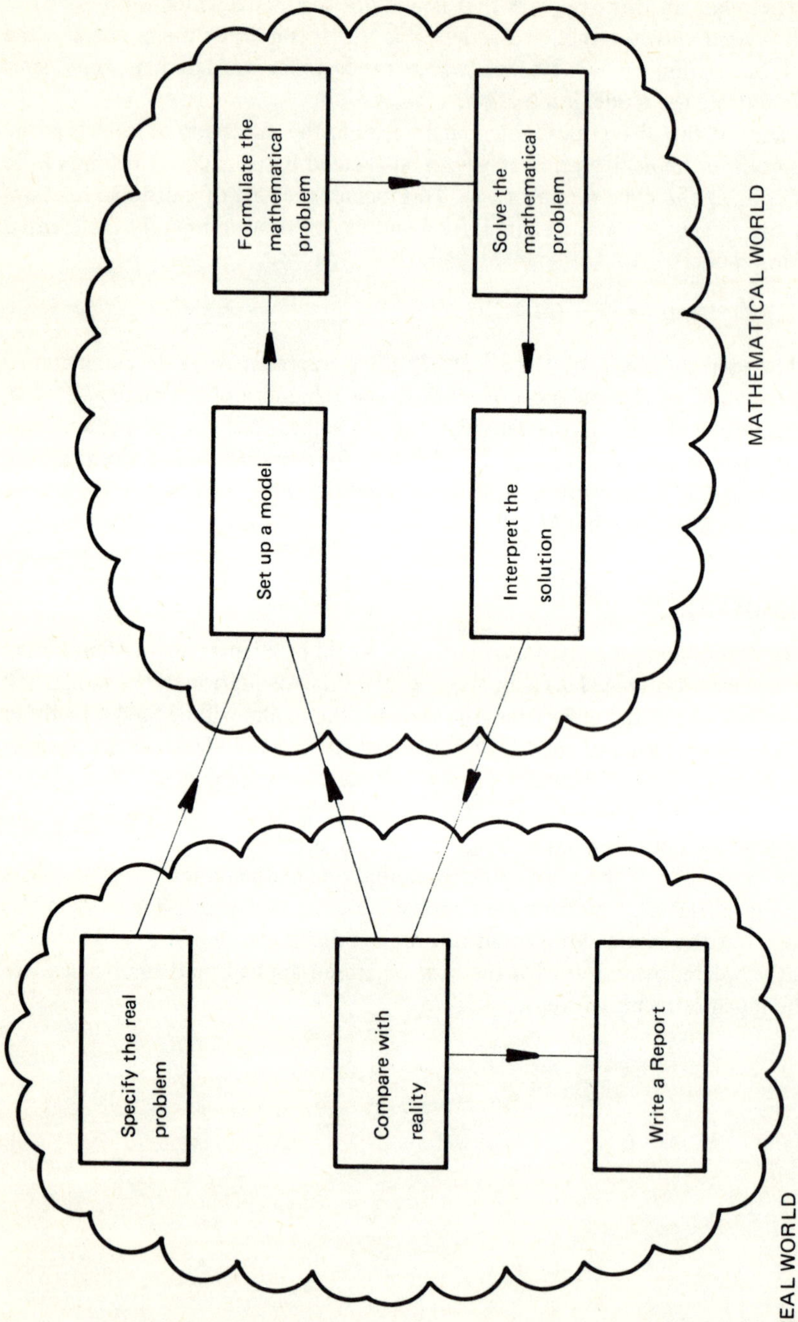

Specify the real problem
Set up a model
Formulate the mathematical problem
Solve the mathematical problem
Interpret the solution
Compare with reality
Write a Report
REAL WORLD
MATHEMATICAL WORLD

Fig. 10.1

Part IV

Problems

INTRODUCTION TO PART IV

This Part comprises 26 problems for which we have *not* given solutions in this book. We hope that you will make a real attempt to solve some of these problems before looking up the solutions. Again, choose the examples that really interest you, as we are sure you will make a better attempt at solving these. Solutions to the problems can be found in the following references:

A *Solving real problems with mathematics,* Vol. 1, CIT Press, 1981.
B *Solving real problems with mathematics,* Vol. 2, CIT Press, 1982.
C *Solving real problems with CSE mathematics,* CIT Press (to be published).
D *Case Studies in Mathematical Modelling,* editors, D. J. G. James and J. J. McDonald, Stanley Thornes, 1981.

The information given after the title of each problem refers to the books listed above and to the original author.

1 FOOTBALL LEAGUE (C: R. Ingledew and R. Biddlecombe)

The football season 1981/82 has seen the introduction of a new points system for the football league tables, namely

3 points – win
1 point – draw
0 points – lose

The previous system differed in having 2 points for a win. Investigate whether this change affects the positions of clubs in the tables. Can you suggest an improved points system? If so, investigate its use and again determine if positions in the table are radically altered.

2 BOOK OF STAMPS (C: N. Green and D. Burghes)

Background

Books of postage stamps are currently produced by the British Post Office for sale in Post Offices and from vending machines. There are books containing the current 1st and 2nd class stamps and ones with a mixture of denominations. At the time of writing there is a 50p book for sale consisting of a selection of stamps to cater for the present 1st and 2nd class postage rates of 14p and 11½p.

Obviously inflation will continually affect both the postage rates and the price of stamp books, and with the advent of a £1 coin (planned for 1983) we can surely expect stamp books with a value of £1, as these would then be easily sold via vending machines.

Problems

(1) Design a 50p book of stamps to cater for the current 1st and 2nd class rates for letters (inland).

(2) Find the values of stamps actually included in a 50p book by the Post Office and compare these values with the ones you chose yourself. Have you any criticisms of the Post Office's choice?

(3) Design a £1 book of stamps, also catering for the current 1st and 2nd class postage rates for letters.

3 LEAPS IN TIME (C: I. Huntley and G. Nellist)

The jingle

> "Thirty days hath September, April, June and November,
> All the rest have thirty-one, excepting February alone,
> With twenty-eight days each common year,
> Leap year coming once in four,
> February then has one day more."

is very well-known in one form or another. Do you know, though, *why* we have a leap year every four years? Or how to tell *which* year is a leap year and which is not?

Go along to the library and see what information you can find about the earth going round the sun, the earth revolving about its own axis, and the number of days in a year. Then try to answer the questions above.

4 MORSE CODE (C: R. Haydock and J. Walton)

Background Information (from *Chambers' Encyclopedia*)

Morse code of dot-and-dash signals in telegraphy was named from S. F. B. Morse who evolved it for his electromagnetic telegraph. It is used for communicating either electrically between stations each equipped with transmitter and receiver connected by wire, or visually (usually by lamp or hand-flag). The problem before Morse was, given the solitary modulation of current 'off' or 'on', then available, to devise at least 26 signals equivalent to the alphabetical letters. He solved it by enlisting the time factor. If a momentary contact gave a dot, a longer contact would produce a dash. The code is evolved from these two simple signs and equally simple spacing. A dash is the length of three dots; the blank interval between dot or dash components of each group (letter) is equal to one dot, that between each letter, to six (in modern usage, five) dots. In coding the alphabet he first exhausted the mathematical possibilities of arranging the dot and dash singly and in groups of two or three at a time, then drew on four-sign combinations. He allotted the briefer signals to the letters most used (in English). All numerals had five-sign groups. Transmitters are essentially switches with spring finger-keys. Receivers in their simplest forms either render the dots and dashes audible on sounders, whence they are read by ear, or print them on a moving paper tape.

A	·-	J	·---	S	···	2	··---
B	···-	K	-·-	T	-	3	···--
C	-·-·	L	·-··	U	··-	4	····-
D	-··	M	--	V	···-	5	·····
E	·	N	-·	W	·--	6	-····
F	··-·	O	---	X	-··-	7	--···
G	--·	P	·--·	Y	-·--	8	---··
H	····	Q	--·-	Z	--··	9	----·
I	··	R	·-·	1	·----	0	-----

The international signals for letters and numerals in the Morse system. The dot lasts about $\frac{1}{26}$th of a second when signalled on a hand telegraph key, and the dash three times as long.

Reprinted with the permission of The Caxton Publishing Co. Ltd.

Problems

We might take it that Morse's aim was to transmit words as quickly as possible, using his dot-dash idea. If so, is this the best code? Or could you suggest any improvements?

Try solving the problem in your own way if you like. You may find it hard to make a start, so here are some suggestions.

(1) Make a list of all the possible code-names using one, two, three and four signs. The start of the list is shown below (with all the two-sign code-names put in for you).

No. of signs

1	2	3	4
	·· I		
	·- A		
	-· N		
	-- M		

Put the letters of the alphabet alongside the corresponding code-names. Which code-names are not used?

If a letter is common, like E, we want to be able to transmit it quickly, but if it is rare, like Z, we don't mind if it takes longer.

(2) Make up a table of how often each letter of the alphabet recurs by taking a page in a book and counting them. The best way to do this is to use a tally table.

Letter	Tally	Frequency
A	~~1111~~ ~~1111~~ 111	13
B	1111	4
C	~~1111~~ 11	7
D	111	3

Now arrange the letters in order with the most common first.

(3) Write down alongside each of the letters how many 'dot-lengths' of time it takes to transmit. Remember that a dash is the length of 3 dots, and the gaps between the signs are the same length as a dot.

Are the most frequent letters always the shortest to transmit? If not, how could you improve the code?

5 BUY OR RENT TV (A: R. Davison and R. Blackford)

A person wishes to acquire a colour television set. The problem is 'Should the set be bought or rented?'

Some of the factors affecting the decision are:

Cost of electricity
Cost of licence
Depreciation of the value of a set
Monthly rental charges.

There are many other factors!

6 SICKLE CELL ANAEMIA (A: J. Walton and R. Davison)

Background

(a) *Genes*

A lot of our characteristics are inherited from our parents, e.g. if both parents have black hair then their children will very probably have black hair too. These characteristics are transmitted in special bundles of chemicals called GENES. Genes occur in pairs – one from each parent.

(b) *Heterozygotes and Homozygotes*

Heter means 'opposite'

Homo means 'the same'

Suppose there are two genes X and Y,

e.g. X could be the gene 'black hair'

Y could be the gene 'blue eyes'.

If a mother and father both have genes X and Y their children could have gene pairs XX, XY, YX or YY. XY and YX are the same; they both mean 'black hair and blue eyes'.

		Mother's genes X	Mother's genes Y
Father's genes	X	XX	XY
	Y	YX	YY

Children with the gene pair XX or YY are called HOMOZYGOTES; those with gene pairs XY are called HETERZYGOTES.

(c) *Genetics and Probability*

One important question which doctors wish to answer is 'What is the probability of a child having a certain gene pair?'

We will look at the genes X and Y; we look at the situation of a particular pair of genes, each of which must be an X gene or a Y gene. The probability of picking gene X at random is p and of picking gene Y is q. (What is the connection between p and q?) This is not the same probability as picking a *person* at random with gene X. Can you think of an example to illustrate this?

The probability a child will have the gene pair XX is –

The probability a child will have gene pair XY is –

The probability a child will have gene pair XY is – .

Can you fill in the blanks?

This problem was first investigated by scientists Hardy and Weinberg who have a law named in their honour.

Hardy–Weinberg Law

The probability of gene X being placed in box 1 is p.

The probability of gene Y being placed in box 2 is q.

So the probability of two X genes being placed in box 1 is $p \times p = p^2$, and the probability of two Y genes being placed in box 2 is $q \times q = q^2$.
The probability of an X gene in box 1 and a Y gene in box 2 is $pq + qp = 2pq$,
e.g. If 3 in 10 genes are X, 7 in 10 are Y, then $p = 0.3$, and $q = 0.7$.
Probability of XX gene pair is $(0.3)^2 = 0.09$.
Probability of YY gene pair is $(0.7)^2 = 0.49$.
Probability of XY gene pair is $2(0.3)(0.7) = 0.42$.
Note that total probability is 1, so that all possible cases have been covered.

Mother's genes

		X p	Y q
Father's genes	X p	p^2	pq
	Y q	qp	q^2

The Hardy–Weinberg law states that if gene X occurs with probability p and gene Y with probability q, then the probability of the gene pairs XX, XY, and YY are p^2, $2pq$ and q^2 respectively.

Problem

There is a certain gene, which we will call X, which results in the disease sickle cell anaemia in people with the gene pair XX. This is caused by a deformation of the oxygen-carrying haemoglobin molecules in the blood, which tend to break down under stress conditions such as even a slightly lowered oxygen concentration in the atmosphere. People with this disease rarely live past the age of 30, and, because of this, the incidence of the disease in the UK is very low, about 0.02% of the population.

There are, however, parts of the world where the disease is much more common. This is because a single X gene, i.e. an XY pair, confers some immunity to salcitarium malaria, itself a potentially fatal disease. Indeed, about 45% of the negro population of Africa carry at least one X gene and about 7% are born with XX. Even among the negro population of the USA, 10% carry either the single or double X gene.

Some Questions

(a) Knowing the proportion of births with sickle cell anaemia, what is the underlying proportion of gene X?
(b) What will be the proportion of gene X in the future? Will it ever die out completely?
(c) How does the incidence of malaria affect the long-term incidence of sickle cell anaemia?

7 ECLIPSE (A: J. Berry and B. Young)

Background information

The sun can be considered as a large sphere of gas, stationary in space. The earth is one of the planets of the solar system (special because we live on it!) moving round the sun, and the path of the earth is an ellipse with the sun at one focus. The eccentricity of the ellipse is very small, so that the earth's orbit is very nearly a circle; the distance of the earth from the sun varies between 146.2×10^8 km and 151.4×10^8 km.

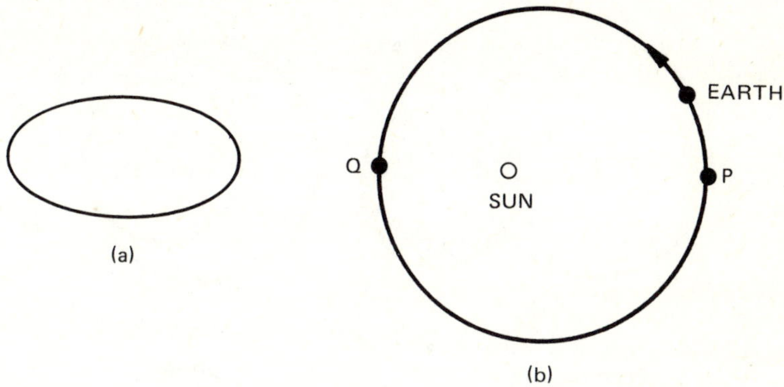

Fig. 7.1

The usual ellipse looks like (a), e.g. like a rugby ball. The earth's orbit, (b), is very nearly circular. The positions of greatest and least distances (points P and Q respectively) are called aphelion and perihelion.

The moon is a satellite of the earth (Fig. 7.2). It revolves around the earth once every 28 days; the path of the moon is not quite circular, its distance from the earth varies from 3.62×10^5 km to 4.03×10^5 km. However, like the motion of the earth around the sun, a circular orbit is a good simple model for the path.

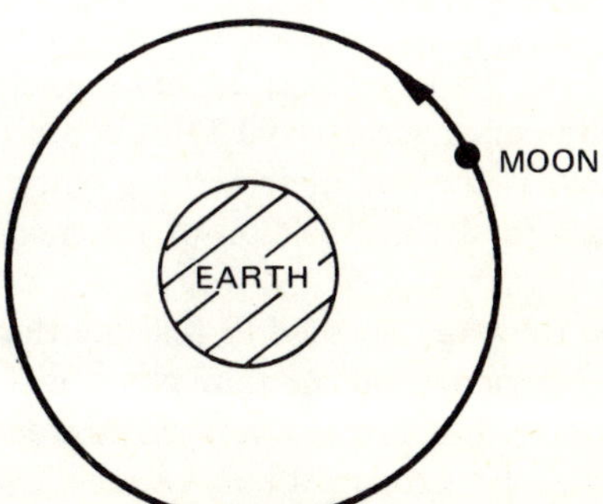

Fig. 7.2

When we look at the moon and the sun from earth we see two discs of very nearly the same diameter. Now when these discs are lined up so that the moon is directly in front of the sun, then the sun almost disappears from view. This phenomenon is called a *solar eclipse.* All that is visible is a black disc and a bright outer ring called the corona. We are very fortunate that nature has made the sun and moon the sizes and distances from the earth that they are, because otherwise we might not see the corona at all. Figure 7.3 shows the four phases of a solar eclipse.

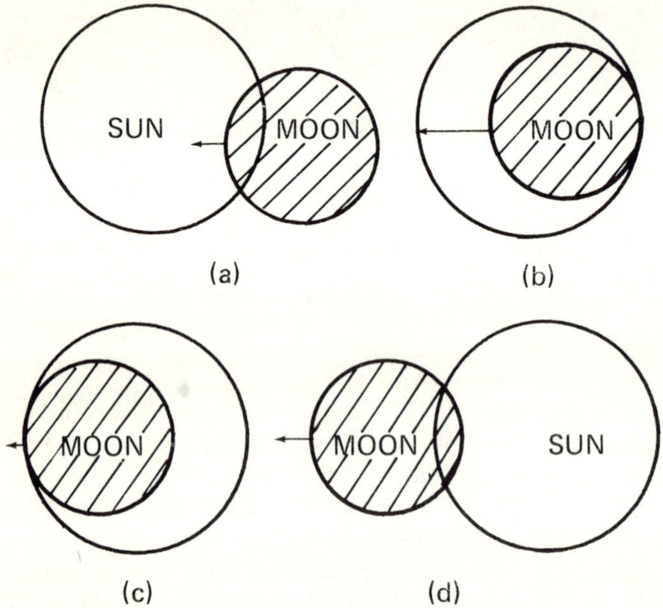

Fig. 7.3

At stage (a), called 'first contact', the moon's disc begins to cross the sun's disc; (b) and (c) are called second and third contact, and phase (d) is called fourth contact which is the end of the eclipse when the moon and sun's discs separate. The arrows show the direction of the moon's motion relative to the sun.

In this exercise the pupil is asked to find the time between (i) the first and fourth contact and (ii) the second and third contact.

The type of eclipse described above is called a *total* eclipse. When it occurs it can be seen along a very narrow path on the earth's surface. This is because the shadow of the moon on the earth's surface is a circle of diameter 269 km (e.g. the distance from London to Sheffield).

Fig. 7.4

Figure 7.4 shows the path of the total eclipse that occurred on 16 February 1980. This path is called the zone of totality. The shadow of the moon travelled along this path, which has a width of approximately 269 km. For a distance of about 3000 km each side of the path the moon passed in front of the sun, but at no time completely hiding it. This called a *partial eclipse* and is illustrated in Fig. 17.5. The further away from the zone of totality you go, then the more of the sun is seen.

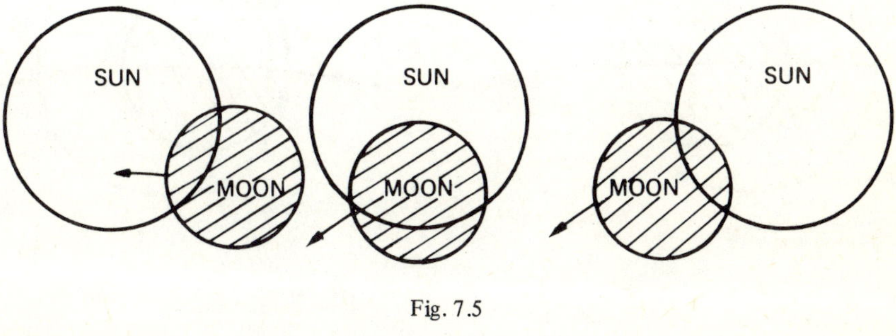

Fig. 7.5

The next total eclipse visible in England occurs in 1999. Let's hope it is not cloudy!

REFERENCES

[1] Muirden, J., *Guide to astronomy*, Pan Books.
[2] Payne-Gaposchkin, C., *Introduction to Astronomy*, University Paperbacks.

Problems

(1) Find the length of time that the moon takes to cross the face of the sun. Figure 7.6 shows the start and end of the period required.

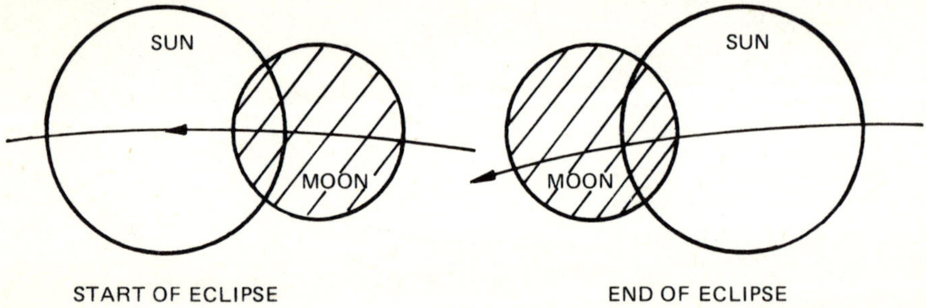

Fig. 7.6

(2) Find the time while the moon is completely in front of the sun (i.e. the total eclipse time) (Fig. 7.7).

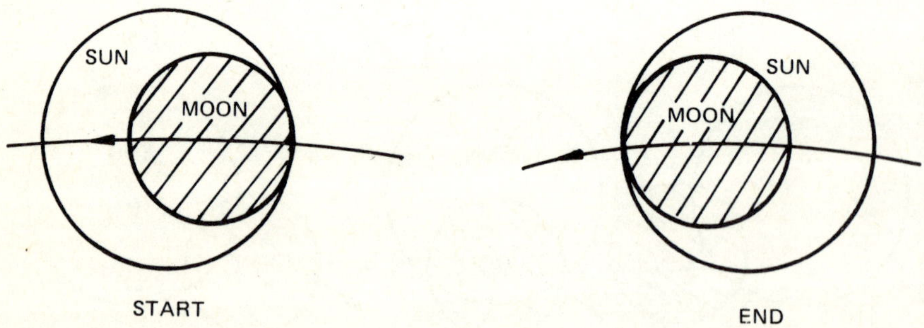

Fig. 7.7

8 RUGBY GOALKICKING (B: R. Ingledew and I. Huntley)

Background Information

When a try has been scored in Rugby, the player taking the kick at goal is allowed to move as far downfield as he wishes from the point where the try was scored to place the ball for the kick, as illustrated in Fig. 8.1. (Some dimensions are given, but the diagram is not to scale.)

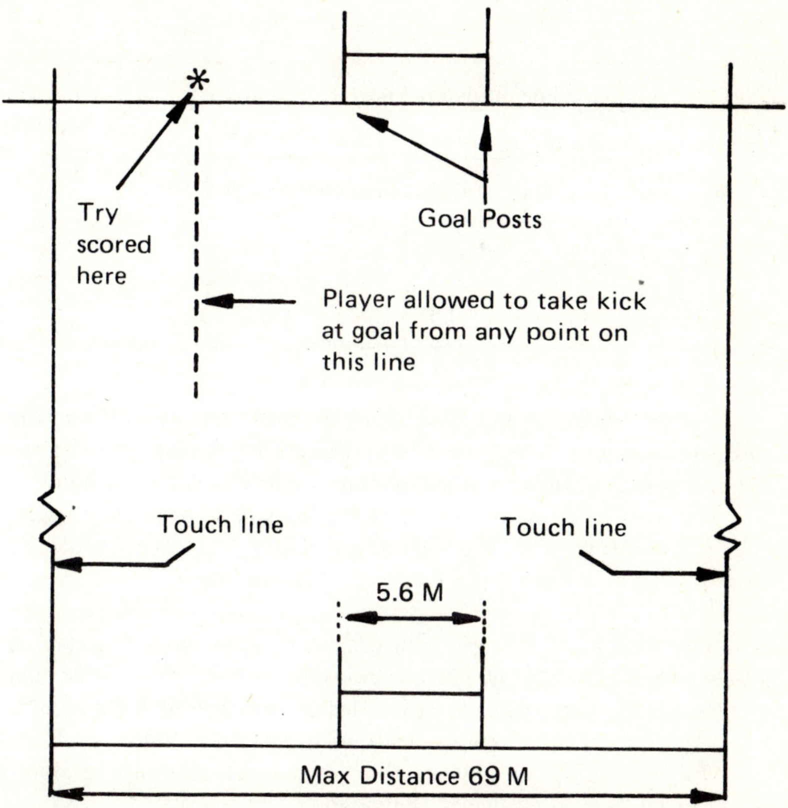

Fig. 8.1

Problem Statement

Where should the ball be placed to maximise the chances of a successful goalkick?

9 FARM (B: J. Berry and B. Young)

(1) Suppose that you are a farmer and have a small plot of area 16 hectares (1 hectare $\simeq$ 2½ acres) on which you want to grow two crops labelled X and Y. The production of each crop requires the following three operations: (i) ploughing, (ii) drilling, and (iii) harvesting. Each of these operations takes a certain amount of time and is carried out at different periods of the year.

If we divide the year into three periods, then the time needed for each operation is shown in the table:

Period	Time needed per hectare in hours						Total time available in hours
	Crop X			Crop Y			
	harvest	plough	drill	harvest	plough	drill	
1	1.5						20
2	•	0.5		2.0			15
3			0.3		0.5		5

The last column in this table shows the total time available for farming this particular plot during each period. The drilling of crop Y can be carried out at that time of the year when the time available is not a constraint.

You want to divide the area between the two crops so as to achieve as large a profit as possible, and you know that crop Y is twice as profitable as crop X. What areas should you use for crop X and crop Y?

(2) A farmer with a much larger farm decides to grow crops X and Y as in problem (1), but he has 160 hectares available. He employs a certain number of men on the farm, and the farmer assigns for each man the number of hours in column 4 (total time available in hours) of the Table in problem 1.

The profit made on growing these crops must take into account the wages to be paid to each man. Suppose that each man costs 15 units of profit. How many men does the farmer need to employ in cultivating this field so as to obtain as large a profit as possible and use all the land available?

10 TITANIC INSURANCE CO. LTD. (B: J. Walton and R. Davison)

The Titanic Insurance Co. Ltd.
Internal memo

From: Managing Director	To: Office Services Manager
Date: 8/12/80	Copies to: All Dept. Heads

I have received numerous complaints that large numbers of our employees are reaching their offices well after 9.00 am due to the inability of the present three lifts to cope with the rush at the start of the day. In the present financial situation it is impossible to consider installing any extra lifts or increasing the capacity of existing ones above the current ten person. Please investigate and let me have some possible solutions to the problem with an indication of their various advantages and disadvantages.

The Titanic Insurance Co. Ltd.
Internal memo

From: Office Services Manager	To: Personal Assistant
Date: 9/12/80	Copies to:

Mary,

Can you find out:

1. How long the lifts take to get between floors and how long they stop for.
2. How many people from each floor use the lift on the way in in the morning.
3. How many people were late this morning?

Frank.

The Titanic Insurance Co. Ltd.
Internal Memo

From: P.A.	To: Office Services Manager
Date: 10/12/80	Copies to:

Frank,

The lifts:

1. Lifts appear to take about 10 seconds between each floor, an extra 15 seconds for each stop and another 5 seconds if the doors have to reopen. It also seems to take about 25 seconds to fill the lift on the ground floor.

2. I have been round and asked, and it looks as if on an average day the numbers are:

Floor	1st	2nd	3rd	4th
No. people	10	30	100	100

3. About 60 people were late today.

Mary

The Titanic Insurance Co. Ltd.
Internal memo

From: Office Services Manager	To: Managing Director
Date:	Copies to:

?

11 STORAGE RACKS (B: B. Young and I. Huntley)

A wine merchant is planning to open a new shop. Behind the shop is a small store-room which he intends to use as a stockroom for the more rapidly moving stock. There is also a larger room 10 m by 6 m by 2 m high which he intends to fit out as a wine cellar, so that he can buy wine relatively cheaply when young and store until it is ready for drinking and hence will fetch a higher price.

The wine cellar will contain racks in which bottles will be stored horizontally so that each bottle may easily be moved from its rack. The racks will be fixed so that the wine may remain undisturbed whilst maturing.

Estimate the maximum number of 70–75 cl bottles that could be stored in this way in the vertically-walled cellar.

12 CALLING ALL NOAHS (B: J. Berry and B. Young)

The burning of fossil fuels adds carbon dioxide to the atmosphere around the earth. This may be partly removed by biological reactions, but the concentration of carbon dioxide is gradually increasing. This increase leads to a rise in the average temperature of the earth. The table below shows this temperature rise over the last 100 years.

Year	Temperature rise of the earth over the 1860 figure (°C).
1880	0.01
1896	0.02
1900	0.03
1910	0.04
1920	0.06
1930	0.08
1940	0.1
1950	0.13
1960	0.18
1970	0.24

If the average temperature of the earth rises by about another 6°C from the present value, this would have a dramatic effect on the polar ice caps, winter temperature, etc. If the ice caps melt there will be massive floods, and a lot of the land mass would be submerged. The UK would disappear except for the tops of the mountains!

Find a model of the above data and use it to predict when the earth's temperature will be 7°C above its 1860 value. Should you start building an ark yet?

13 THE DEFENCE OF JEDESLAND (B: R. Blackford and D. Burghes)

Background Information

The cold war between Jedesland and Andersland has been hotting up. In order to assess the present strength and deployment of Jedesland forces, Andersland has been sending automatic (pilotless) spy-planes at high altitudes over Jedesland.

The Premier of Jedesland is determined that this will not continue and has managed to purchase from the weapons black market his first anti-aircraft missile, FRED 1.

The guidance system of FRED 1 is such that it locks onto the heat source of the aircraft (i.e. engine), and it will always move in the direction of this source until either it hits the aircraft or it runs out of fuel.

Problem

The Ministry of Defence has just informed the Premier that a suspected spy-plane from Andersland has been spotted and will soon be flying directly over the Jedesland missile base at a constant speed (estimated to be 1500 ms^{-1}) and at an estimated height of 4000 m.

The plane still has about 8000 m to travel over Jedesland national territory, and the speed of FRED 1 is 2000 ms^{-1}.

What advice would you give to the Premier? It should be noted that if the plane is to be shot down, it must be accomplished over the Jedesland territory, and not outside it, which would cause international opinion to move against Jedesland.

14 COVER UP (B: J. Walton and R. Davison)

A manufacturer of picture mirrors wishes to cover them in clear plastic wrapping film which is then heat-sealed. The mirrors come in two sizes, 20 cm by 20 cm and 30 cm by 40 cm and are 4 mm thick.

Work out how you can wrap these mirrors using the minimum amount of plastic film.

15 DIPSTICK (B: I. Huntley and R. Ingledew)

Central heating tank

Many oil-fired central heating systems store the fuel in a rectangular oil tank which has a dipstick provided, marked in gallons, to enable the user to see when the level is low and to reorder. Imagine that you have just moved into a new house, which is provided with such a system, but that the dipstick has been lost! Can you make a new one from a handy length of steel rod?

Underground tank

Imagine now a very similar situation: you have just bought a disused garage, which you know has an underground cylindrical tank, but no dipstick! You

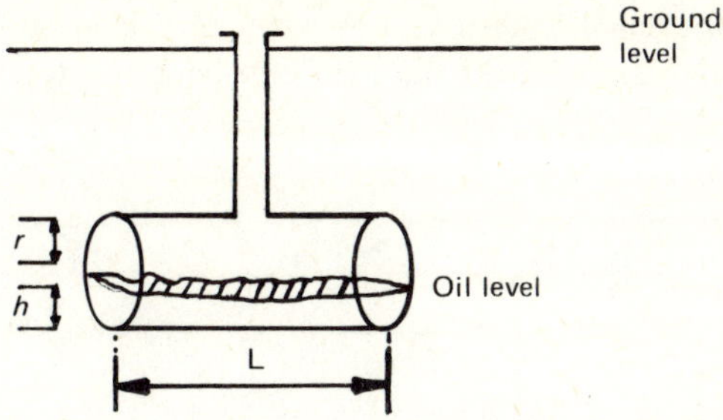

realise quite quickly how to find the radius by dipping the top and bottom of the cylinder (how would *you* do it?), and hence know that $r = 2$ m. You still don't know the length L, but know that the level rises from 25 cm to 110 cm when you add 3000 gallons to the tank. Can you make a dipstick this time?

Underground tank revisited

The man who told you the underground tank was a circular cylinder has just apologised for making a mistake. He was hunting through his papers, and found

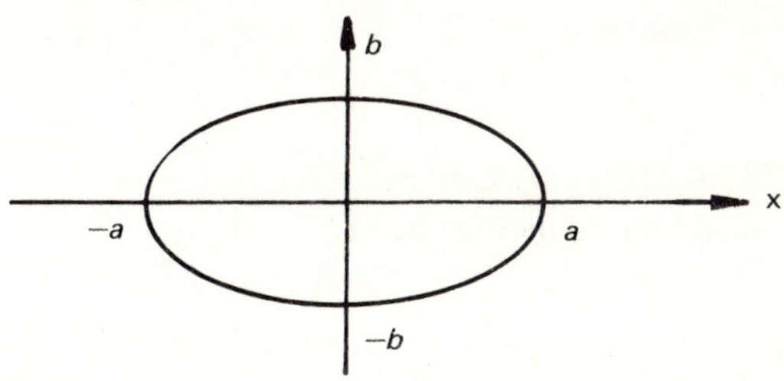

a document that says the cross-section is really an ellipse as shown and that the tank is 5 m long! Fortunately you hadn't quite calibrated your dipstick. Can you do so, now?

16 CAR PARK (B: A. Hunter and B. Young)

Car parking is permitted free along a straight length of road overlooking the sea at a seaside town. At present cars may be parked at any point along the road where there is sufficient space, but must be parallel to the road edge.

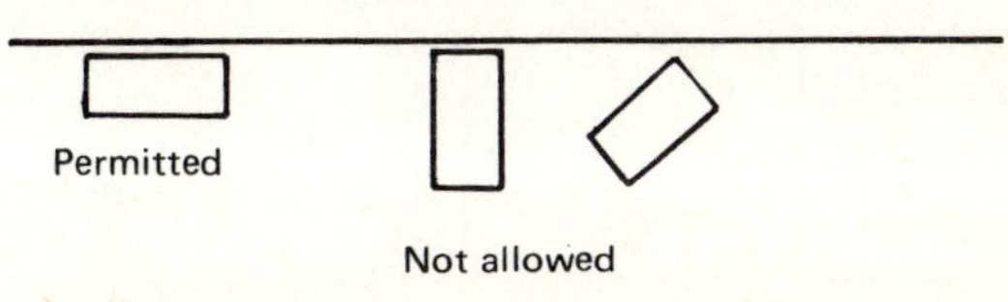

Present System

The local council is keen that as many cars as possible can be parked there. It is considering marking off the road into parking bays, each long enough for one car.

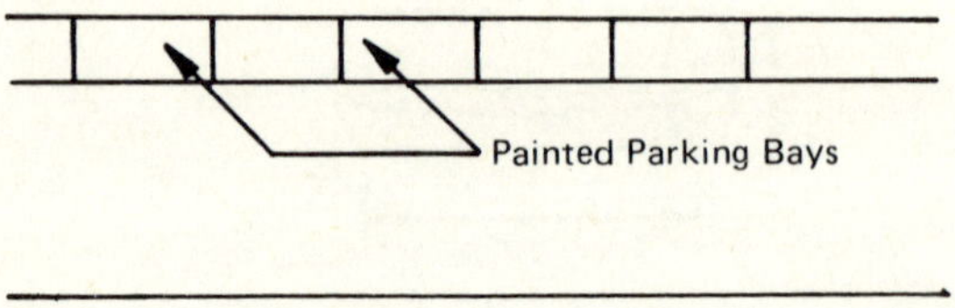

Suggested New System

Will there be an increase in the number of cars that can be parked? If so what improvement would it be reasonable for the council to expect?

Obviously the maximum number of cars in the new system is the same as the number of bays. The problem is to find out how many cars could be parked using the *present* system. (This was where each driver could choose to park at any available point along the road.)

Would your conclusions be the same if cars were parked at right-angles to the kerb?

Data: Typical length of a parking bay is 5 m.
Typical width of a parking bay is 2.3 m.

17 TRANSFER FEES (B: G. Nellist and R. Blackford)

The factors that affect the performance of a league football team, and its final position at the end of the season, are clearly many and varied. However, one of the most important of these must certainly be the skill and calibre of the eleven players on the field. In an effort presumably to improve their performance, many teams take to the market place to buy and sell players for ever escalating transfer fees. Are these fees justified? Do those clubs who spend most, reap the benefits of larger crowds and better results?

PLAYFAIR FOOTBALL ANNUAL 1980–81 presents a great deal of information relevant to these questions, in particular the following tables will be found to be useful.

TRANSFER TRAIL

(players with British Clubs)

First four-figure transaction
Alf Common: Sunderland to Middlesborough for £1,000 in February 1905.

First five-figure transaction
David Jack: Bolton Wanderers to Arsenal for £10,340 in October 1928.

First six-figure transaction
Alan Ball: Blackpool to Everton £112,000, August 1966.

First £200,000 transaction
Martin Peters: West Ham United to Tottenham Hotspur £200,000, March 1970.

First seven-figure transaction
Trevor Francis: Birmingham City to Nottingham Forest £1,000,000, February 1979.
Andy Gray: Aston Villa to Wolverhampton Wanderers £1,469,000, September 1979.
Steve Daley: Wolverhampton Wanderers to Manchester City £1,450,000, September 1979.
Clive Allen: Queen's Park Rangers to Arsenal £1,250,000, June 1980.
Trevor Francis: Birmingham City to Nottingham Forest £1,000,000, February 1979.
Kevin Reeves: Norwich City to Manchester City £1,000,000, March 1980.
Steve Archibald: Aberdeen to Tottenham Hotspur £1,000,000, May 1980.
Mick Robinson: Preston North End to Manchester City £756,000, July 1979.
Ray Wilkins: Chelsea to Manchester United £750,000, August 1979.
Mike Flanagan: Charlton Athletic to Crystal Palace £722,000 August 1979.
Peter Barnes: Manchester City to West Bromwich Albion £650,000, July 1979.
John Gidman: Aston Villa to Everton £650,000, October 1979.
Phil Parkes: Queens Park Rangers to West Ham United £565,000, February 1979.
David Mills: Middlesbrough to West Bromwich Albion £516,000, January 1979.
Asa Hartford: Manchester City to Nottingham Forest £500,000, July 1979.
Asa Hartford: Nottingham Forest to Everton £500,000, August 1979.
Gerry Francis: Queen's Park Rangers to Crystal Palace £465,000, July 1979.
Gordon McQueen: Leeds United to Manchester United £450,000, February 1978.
Brian Talbot: Ipswich Town to Arsenal £450,000, January 1979.
Garry Owen: Manchester City to West Bromwich Albion £450,000, June 1979.
Kenny Dalglish: Celtic to Liverpool £440,000, May 1979.
Osvaldo Ardiles: Huracan to T ttenham Hotspur £400,000, July 1978.
Charlie George: Derby County to Southampton £400,000, December 1978.
Tony Currie: Leeds United to Queen's Park Rangers £400,000, August 1979.
Frank Gray: Leeds United to Nottingham Forest £400,000, August 1979.
John Deehan: Aston Villa to West Bromwich Albion £400,000, September 1979.
Gordon Smith: Rangers to Brighton & Hove Albion £400,000, May 1980.
Ray Stewart: Dundee United to West Ham United £400,000, September 1979.
Dave Thomas: Everton to Wolverhampton Wanderers £400,000, October 1979.
Dave Swindlehurst: Crystal Palace to Derby County £400,000, April 1980.
Alex Sabella: Sheffield United to Leeds United £400,000, May 1980.
Alan Curtis: Swansea City to Leeds United £375,000, May 1979.
Irving Nattrass: Newcastle United to Middlesbrough £375,000, August 1979.
Kevin Hird: Blackburn Rovers to Leeds United £360,000, March 1979.
Graeme Souness: Middlesbrough to Liverpool £352,000, January 1978.
Bob Latchford: Birmingham City to Everton £350,000, February 1974.
Joe Jordan: Leeds United to Manchester United £350,000, January 1978.
Paul Futcher: Luton Town to Manchester City £350,000, June 1978.

Richardo Villa: Racing Club to Tottenham Hotspur £350,000, July 1978.
Len Cantello: West Bromwich Albion to Bolton Wanderers £350,000, June 1979.
Garry Collier: Bristol City to Coventry City £350,000, July 1979.
Brian Greenhoff: Manchester United to Leeds United £350,000, August 1979.
Alan Biley: Cambridge United to Derby County £350,000, January 1980.
Malcolm Macdonald: Newcastle United to Arsenal £333,333, August 1976.
Richard Money: Fulham to Liverpool £333,000, May 1980.
Paul Hart: Blackpool to Leeds United £330,000, March 1978.
Colin Todd: Derby County to Everton £330,000, September 1978.
Peter Shilton: Leicester City to Stoke City £325,000, November 1974.
Mick Walsh: Blackpool to Everton £325,000, August 1978.
Claudio Marangoni: San Lorenzo (Argentina) to Sunderland £320,000, December 1979.
Martin Dobson: Burnley to Everton £300,000, August 1974.
Leighton James: Burnley to Derby County £300,000, December 1975.
Mike Channon: Southampton to Manchester City £300,000, July 1977.
Alan Kennedy: Newcastle United to Liverpool £300,000, August 1978.
Mike Thomas: Wrexham to Manchester United £300,000, November 1978.
Frank McGarvey: St Mirren to Liverpool £300,000, May 1979.
John Gregory: Aston Villa to Brighton & Hove Albion £300,000, July 1979.
Bobby Shinton: Wrexham to Manchester City £300,000, July 1979.
Garry Stanley: Chelsea to Everton £300,000, August 1979.
Terry Yorath: Coventry City to Tottenham Hotspur £300,000, August 1979.
David Geddis: Ipswich Town to Aston Villa £300,000, September 1979.
Steve Wicks: Derby County to Queen's Park Rangers £300,000, September 1979.
Barry Powell: Coventry City to Derby County £300,000, October 1979.
Stan Cummins: Middlesbrough to Sunderland £300,000, November 1979.
Ian Rush: Chester to Liverpool £300,000, May 1980.
Dave Watson: Sunderland to Manchester City £280,000, June 1975.
Derek Hales: Charlton Athletic to Derby County £280,000, December 1976.
Peter Eastoe: Queen's Park Rangers to Everton £280,000, March 1979.
Mick Walsh: Everton to Queen's Park Rangers £280,000, March 1979
Peter Osgood: Chelsea to Southampton £275,000, March 1974.
Dennis Tueart: Sunderland to Manchester City £275,000, March 1974.
Steve Wicks: Chelsea to Derby County £275,000, January 1979.
Des Bremner: Hibernian to Aston Villa £275,000, September 1979.
Frank McGarvey: Liverpool to Celtic £275,000, March 1980.
Peter Shilton: Stoke City to Nottingham Forest £270,000, September 1977.
Asa Hartford : West Bromwich Albion to Manchester City £250,000, August 1974.
Tommy Craig: Newcastle United to Aston Villa £250,000, January 1978.
Gordon Hill: Manchester United to Derby County £250,000, April 1978.
Neil McNab: Tottenham Hotspur to Bolton Wanderers £250,000, November 1978.
David Jones: Everton to Coventry City £250,000, June 1979.
Stephen Mackenzie: Crystal Palace to Manchester City £250,000, July 1979.
Chris Woods: Nottingham Forest to Queen's Park Rangers £250,000, July 1979.
Tony Fitzpatrick: St Mirren to Bristol City £250,000, August 1979.
Mike Pejic: Everton to Aston Villa £250,000, September 1979.
Colin Todd: Everton to Birmingham City £250,000, September 1979.
Stan Bowles: Queen's Park Rangers to Nottingham Forest £250,000, December 1979.
Roger Van Gool: FC Cologne to Ipswich Town £250,000, March 1980.

ATTENDANCES

ATTENDANCES UP

Attendances at Football League matches in 1979-80 were slightly up overall compared with the previous season, but the interesting fact is that whereas the First and Second Divisions were down on last season, there was a sufficient increase in those for Third and Fourth Divisions to outweigh that. Does this mean that football in these divisions is comparatively entertaining? There was no question about the best-supported club – Manchester United once again – but even Rochdale, who finished 92nd out of 92 showed some improvement! The divisional totals, with last year's figures in brackets, were:

Division One: 12,163,002 (12,704,549)
Division Two: 6,112,025 (6,153,223)
Division Three: 3,999,328 (3,374,558)
Division Four: 2,349,620 (2,308,297)

DIVISION ONE

	League position	Average 1979-80	Average 1978-79	Difference
Arsenal	4	33,596	36,371	−2,775
Aston Villa	7	27,978	32,838	−4,860
Bolton Wanderers	22	16,353	24,772	−8,419
Brighton & Hove Albion	16	24,745	22,145	+2,600
Bristol City	20	18,932	22,306	−3,374
Coventry City	15	19,315	22,637	−3,322
Crystal Palace	13	29,794	23,294	+6,500
Derby County	21	19,904	21,555	−1,651
Everton	19	28,711	35,456	−6,745
Ipswich Town	3	21,620	21,673	−53
Leeds United	11	22,788	27,633	−4,845
Liverpool	1	44,586	46,406	−1,820
Manchester City	17	35,272	36,203	−931
Manchester United	2	51,608	46,430	+5,178
Middlesbrough	9	18,739	18,459	+280
Norwich City	12	17,225	17,874	−649
Nottingham Forest	5	26,350	29,587	−3,237
Southampton	8	21,335	21,330	+5
Stoke City	18	20,176	19,125	+1,051
Tottenham Hotspur	14	32,018	34,902	−2,884
West Bromwich Albion	10	22,418	26,517	−4,099
Wolverhampton W.	6	25,731	20,796	+4,935

18 X-RAYS (B: S. Pirie and A. Hunter)

Background

X-rays are produced when high-speed electrons strike a target and are rapidly decelerated. 99% of the energy emitted is dissipated in heat and 1% in X-rays. These X-rays penetrate the object they are aimed at with varying degrees of success depending on the voltage used and the thickness and density of the object. The rays which do penetrate the object cause a reaction to take place in the emulsion of the photographic plate, and an X-ray image is produced as we know it.

The radiographer has various choices open to him. He can select

the voltage (known as the kVp – from kilovolt peak – the penetrating power depends on the peak voltage value not the average),
the current (known as the mA – from milliampere),
the size of the source from which X-rays are emitted (usually 1 mm^2 or 2 mm^2),
the distance of the source from the body (a calibrated scale on the machine, known as the FFD from Film-Focus Distance),
the time of exposure (known as s from second).

The current per second is called the mAs.
These choices are interdependent according to the law which governs all use of X-rays:

$$\text{mAs} = c\frac{(\text{FFD})^2}{(\text{kVp})^2} \quad \text{where } c \text{ is a constant}$$

restricted by the available machines and films in the department.

The radiographer's selection of values depends partly on experience and partly on mathematics. Two factors which influence the choice are the 'geometric unsharpness' (Ug), which is blurring of the image due to penumbra caused by a non-point source, and 'movement unsharpness' (Um) which is blurring of the image due to movement by the patient.

Two further constraints he has to bear in mind are:

(i) the dose of radiation given to the patient, which can be cumulative: a FFD of 100 cm is considered the minimum safe distance from the radiation point of view, and

(ii) the heat produced must be below the manufacturer's permitted maximum to avoid overloading and thus damaging the X-ray machine. The heat emitted by a single-shot X-ray is given by H.U. = kVp × mAs where H.U. are the heat units.

2 Problem

The radiographer's list on 5 January, 1981 reads:

BROWN, Mary, born 12.5.42, suspected fracture of forearm
PRICE, Amanda, born 4.3.80, broken shinbone
SMITH, Peter, born 24.12.38, thoracic spinal examination
JONES, Andrew, born 7.9.11, chest examination
ROBINSON, Elizabeth, born 29.6.61, angiogram.

Thoracic spinal examination involves an X-ray of that part of the spine behind the chest.

An angiogram is a series of 30 X-ray shots of the blood vessels as dye is dispersed through the body.

Determine the conditions appropriate for each of the patients.

3 Additional information

There are 3 X-ray machines:

Number	Source	mA	Anode heat storage capacity
1	1 mm^2	25 to 250	75000
2	1 mm^2	100 to 200	25000
3	2 mm^2	100 to 600	175000

m A is selected from discrete values by pushing the relevant button.
s is selected by a pointer on a continuous range, points discernible to 2 decimal places on machines 1 and 2, and 3 decimal places on machine 3.
FFD is measurable to the nearest whole centimetre.
The KVp is chosen by experience. The necessary values are given below:

for forearm	55 kVp
for shinbone	50 kVp
for thoracic spine	80 kVp
for chest	60 kVp
for angiogram	80 kVp

The mAs gives the density of the image and is governed by the film and intensifying screens in use.

There are 4 types of film and screen available to this radiographer:

A – fast film and high definition screens	mAs 20
B – fast film and fast intensifying screens	mAs 4
C – fast film and fast intensifying screens	mAs 10
D – fast film and high intensifying screens	mAs 75
for angiogram use film	mAs 60

All mAs values given on the films apply to a FFD of 100 cm since this is the standard distance used by radiographers.

19 CROP YIELD (D: A. Barnes)

Every time a farmer, vegetable grower or fruit grower, or even a gardener, decides to grow a crop he must decide how many plants are to be planted on a given area of ground surface. Too few plants will mean that, although the weight of crop from each plant may be large, the yield from a given area of ground surface may not be very high. Too high a plant population density can lead to weak and spindly plants, and the parts of value, e.g. fruits, roots, grain, etc. will be reduced in size, and possibly in number and quality as well. A certain reduction in the value of the crop from each plant can be tolerated because of the compensating effects on yield per unit area of high plant numbers. However, at very high plant numbers yield per unit area can be reduced because of the adverse consequences of small and weak plants. Excessively large plant numbers also tend to increase the chances of disease and pest problems, and can generally increase management problems and costs.

Very often the choice of the optimal number of plants per unit area of ground surface does not present a difficult problem for the farmer or grower. If a well-tried and trusted crop species is grown, the optimal planting density may be well-documented, although examples exist where detailed experimental investigations have shown that substantial yield increases can be obtained by modifying conventionally accepted planting densities. If a crop is being introduced to a region for the first time, or if a new variety of a species is developed or cultural practices are changed, the optimal planting density may not be known. Thus, there can be very important practical reasons for investigating the relationship between crop yield and planting density.

Besides the direct practical needs to investigate the relationship between crop yield and number of plants per unit area of ground surface, occasions often arise in agricultural research when it is necessary to determine the biological basis for the relationship between crop yield and planting density. Because of the biological complexity of the plant (e.g. fruit, grain, roots, etc.) it is simplest to consider the biological basis for this relationship by taking yield to mean all parts of all plants within a crop rather than just the economically valuable components.

Further, the way in which the plants within a crop increase in size from the time of planting to harvest is the fundamental biological process which determines the relationship between crop yield and planting density. After planting, or after seedling emergence if seed are sown, growth of the individual small plants is not influenced by the presence of neighbouring plants and the rate at which their weights increase is approximately proportional to their weights. However, as plants increase in size, the leaves of neighbouring plants begin to overlap; the time at which this process begins and the extent of leaf overlap which occurs will depend upon how many plants are grown per unit area of ground surface. The overlapping of leaves implies partial shading of some plants by others. Except for cases of extremely low plant densities, the whole of the planted area will be covered eventually by leaves, and all plants in the crop will be shaded to a greater or lesser extent.

This process of ever-increasing overlap of leaves and partial shading of neighbouring plants reduces the average intensity of light falling on leaves and leads to reduced levels of photosynthesis, and consequently to reduced plant growth. This process is commonly referred to as plants 'competing' with each other for light. Ultimately, 'competition' for light becomes so intense that the level of photosynthetic activity is only sufficient to replace losses due to leaves dying, etc. and growth ceases completely. An analogous process to competition for light can operate below ground with roots of different plants competing for water and soil nutrients.

The modelling problem is to construct a model to describe the yield (total plant material) of a crop at any time during the period from planting to harvest and to include in this model the effect on yield of differing numbers of plants per unit area. The model should, as far as possible, depend on parameters or terms which have some biological interpretation. Since the main purpose of the model would be to derive from experimental data estimates of those biological parameters or terms which are important in determining the intensity of competition, the smaller the number of parameters or terms consistent with describing the processes involved the better.

20 HYDROELECTRIC POWER GENERATION SYSTEM (D: D. J. G. James)

In places such as the North of Scotland electricity is generated by hydroelectric means. Despite the high average annual rainfall, the topography of the terrain is such that natural catchment areas tend to be small. In order to make full use of the distributive nature of the catchment areas storage dams are arranged at various levels in the hills and operate by gravity. As well as providing a convenient local storage, this network of dams is used to supply water to the lowest dam, which supplies the flow to drive the turbines and so generates the electric power. An important requirement is to maintain the water level in the lowest dam at an appropriate height to ensure that electric generation demand, which is continuously changing with time, is met. Obviously, because of the remoteness of some of the supply dams, it is desirable that the water supply to the lowest dam be controlled automatically and that the mode of control is capable of responding very quickly to demand changes. The controller should also be capable of taking account of disturbances such as direct inflow to the lowest dam from uncontrolled sources, for example, sudden heavy rainfall or a subsidiary stream.

In order to clarify our ideas regarding the mathematical problems involved consider the two-vessel system in cascade, as shown in Fig. 20.1 where each vessel is assumed to have a uniform cross-sectional area.

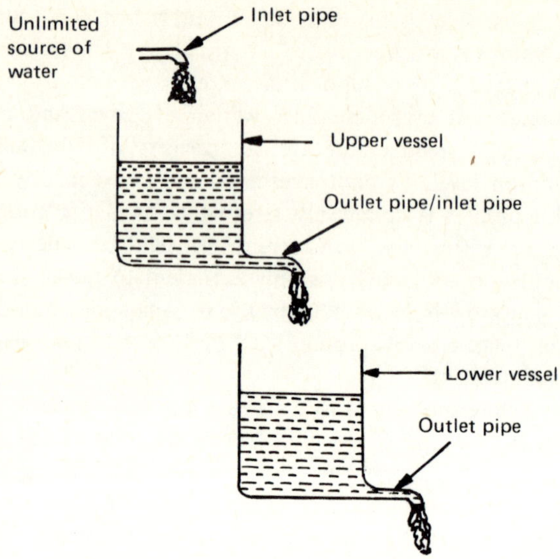

Fig. 20.1 – Two vessels in cascade.

Set up a mathematical model of this two-vessel system and discuss how you would control the inflow to the upper vessel in order to maintain the output flow of the lower vessel at its desired demand value. The amount of water flowing through the inlet and outlet pipes may be controlled by introducing valves. You are required to control the inflow of water in such a way that it can take account of changes in the demand. Also, the controlled system must be capable of reacting to unpredicatable disturbances such as a leakage, or an additional uncontrolled inflow, in one, or both, of the vessels.

21 SPEED-WOBBLE IN MOTOR CYCLES (D: K. Oke)

Many of us have experienced wobbling of the front wheel, felt through the handle-bar, of a motorcycle (or of an ordinary bicycle) when travelling at certain speeds. What causes this wobble, or oscillation, of the steered wheel?

The wobbling phenomenon is not confined to motorcycles and bicycles but it is also known to occur in the front wheels of cars, supermarket or tea trolleys, and in aircraft nose-wheels. These wide-ranging situations all have something in common, namely that the steered wheel is designed as a castor. A castor is defined as a steered rolling wheel, whose point of contact with the ground lies behind the point of intersection of the steering axis and the ground. Figure 21.1 illustrates the basic configuration of two typical castors (not drawn to scale).

Note that the rake angle of a supermarket trolley castor is 0°, that is, the steering axis is vertical. Clearly a number of other important features have to be considered in any attempt to understand how castors oscillate.

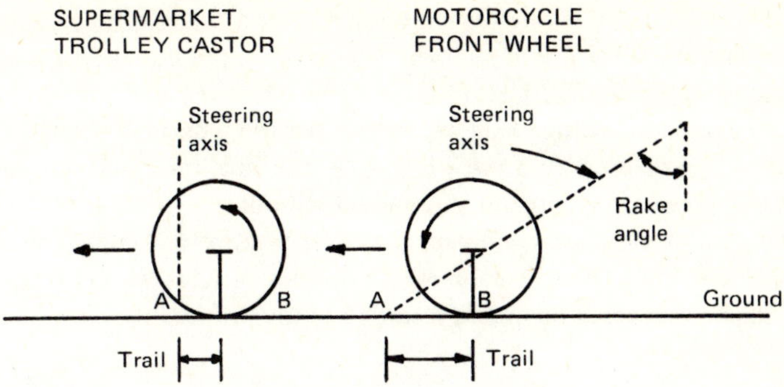

Fig. 21.1 – Typical castors.

For motorcycles, the tyre of the steered wheel and also the suspension in the front forks will obviously affect steering stability. Consequently, there will not be hard (point) contact between the wheel and ground. The area of the 'contact patch' depends on tyre pressure, forward speed of the motorcycle, and whether the cycle is banking on a bend. The flexibility of the tyre also permits lateral movement of the wheel without slipping.

It has been found in practice that the front wheel, even in the 'wheel locked' case (i.e. brakes jammed on hard), can in the case of motorcycles oscillate with a frequency somewhere in the range of 6–8 Hz. In stable cases (which hopefully form the vast majority!), these oscillations rapidly decay to zero.

The problem, then, is to formulate a model which explains some or all of these observations.

Data

Typical motorcycle values, for front wheel oscillations of 6–8 Hz

Speed of motorcycle = 30 ms^{-1}

Moment of inertia of wheel about steering axis = 0.27 kg m^2

Trail = 0.12 m

Coefficient of friction between wheel and ground = 1

Normal reaction between wheel and ground = 700 N

Angular velocity of front wheel about steering axis = ± 12 rad s^{-1} when wheel not turned, i.e. when turning angle is 0.

22 THE NEWS-VENDOR'S DILEMMA (D: R. Croasdale)

'This time last week people were clamouring to buy an evening paper and I'd sold out. So I started ordering more. Now I'm stuck with a pile of papers that nobody wants. I wish I knew how many to stock.'

(Despondent news-vendor)

Imagine that you are taking over the running of a small kiosk retailing sweets, tobacco, and the local evening paper. The paper is published as a single edition every day except Sunday. The question to be answered is 'How many copies of the paper should you order each day?'. This has to be decided in advance because your order is delivered by a van which calls just once each day, and you have no chance of getting more copies if you should sell out.

The printer supplies your papers at a fixed price per copy, and the selling price is fixed too. There is a rebate for copies unsold, whereby some of the outlay can be recovered on copies returned to the printer.

23 MODELLING THE TRUE COST OF A MORTGAGE (D: M. Cross)

Imagine the following scenario some time in the mid-1970s. The manager's office of a well-known building society contains three people Mr and Mrs Smith and the manager himself, Mr Loansome. As we join them the following (typical) conversation is taking place.

Mr Smith: We would like to buy a house, but don't really know much about the availability of mortgages or how much we can borrow. Can you help us?

Mr Loansome: Since you have saved with us for some time, Mr Smith, there should be no trouble in arranging a mortgage for you. The questions you have to consider are what size mortgage you need and which type of mortgage you require.

Mr Smith: We'd obviously like to borrow as much as possible, and I'm afraid we know very little about the different types of mortgage.

Mr Loansome: Let's deal with the amount you wish to borrow first. We normally allow married customers to borrow about two and a half times their joint income. Can you tell me your incomes?

Mr Smith: We have a small child so my wife doesn't work and I earn around £4000 a year as a teacher.

Mr Loansome: So that means you can borrow about £10 000, and since most mortgages are 80% of the house price, you should be looking for a house at around £12 500.

Let us now turn to the type of mortgage. Basically, there are two kinds of mortgage. In the conventional mortgage you borrow a sum at an annual interest (e.g. 10% at present) over a specified term, usually 25 years. We then evaluate a monthly payment in which you pay off a combination of interest and capital. Obviously, at the beginning you are paying interest whilst at the end it is mostly capital. You also have to take out a basic life insurance policy to cover the mortgage in the event of your death, which costs about 35p/month/£1000 borrowed. Thus, on a £10 000 loan your gross monthly payment comes to around £100. However, to help encourage buyers the government gives full tax relief (e.g. 35%) on the mortgage interest payments and 25% on insurance policies. In the early stages this reduces your net monthly payment to about £65.

Turning now to the second kind of mortgage – the endowment. Here you still borrow a sum of money over a specified term, but at a slightly higher interest rate (e.g. 10¼% at present). However, you don't pay back the capital directly. Instead, you pay the interest on the loan and take out an insurance policy which matures either at the end of the specified term or on the death of either of you. When the policy matures it pays off the loan and should leave around 30% of the capital as a profit. Your monthly payment, therefore, has two components, consisting of the interest on the borrowed capital plus the insurance policy at about £1.85/month/£1000 borrowed. This makes your gross monthly payment around £105 although, of course, you still get tax relief as for the conventional mortgage which reduces your net payment to about £70. As you can see, there are two main noticeable differences between the mortgage schemes:

(a) the gross cost of the conventional mortgage is smaller, and
(b) at the end of the term the endowment mortgages leaves you a nice nest egg.

Our friends may now respond as follows.

Mr Smith: My first reaction is that the monthly payment seems quite high for both schemes, though I can now see why it must be, for us to earn 10½% interest on our savings.

Mr Loansome: Yes, the cost appears high, but you must remember your income is likely to increase by at least the annual rate of inflation whilst the payments stay the same, so that after a couple of years the cost becomes relatively small.

Mr Smith: Yes, I suppose so. But which of these mortgages schemes is the best.

Yes – which is the best?

24 SEARCHING FOR GAS LEAKS (D: A. Millington)

The following situation is one which is known to occur. A slight gas leak exists in a domestic gas supply between the meter and the various gas appliances in the house. The pipe above ground is inspected and no leak is detected. One or two floorboards are removed in a further attempt to locate the leak below ground, but without success. At this stage either more floorboards, etc. must be removed in the hope that the leak will be found, or the search abandoned and complete new piping installed.

The problem, therefore, is as follows. A leak exists in a length of pipe, and must be mended either by locating it and repairing or by replacing the piping. The cost of repairing a leak, once located, is small, but it is by no means certain that a leak which does exist in a piece of exposed piping will actually be located by the usual test method. (A typical success rate is 90%, and somewhat less for underground pipe.) In other words, even if every centimetre of pipe is exposed and tested, there is still a chance that no leak will be located.

In these circumstances the Gas Board management sees it appropriate to lay down general guidelines to service engineers in searching for leaks, in particular indicating the stage at which the search should be abandoned.

25 SELLING HOT DOGS (D: A. Rothery)

At a college a hot-dog stall operates over a half-hour period during the lunch interval. The stall operator wishes to run his stall in the most profitable manner, and to this end he has amassed the following details which he considers to have some bearing on the situation.

The rolls used for the hot-dogs cost 50p a dozen, the sausages cost 60p a pound (8 sausages), and the onions about 30p per pound. Each hot-dog needs one roll, one sausage, and ½ oz of onions. In addition about £1's worth of sauce is used over a lunch time. Rolls left over are bought by the staff at 2p each.

The cooking is done using gas cylinders, and about £8's worth of fuel is used over a 5-day week. The cost of equipping the stall is about £5 per week.

Laundry charges amount to about 80p for each person employed plus £1.40 for towels, etc.

It takes about 25 to 35 seconds for an assistant to serve a customer, depending on whether or not the customer requires change. It is thought that about 50% of customers give the correct money.

Table 1 – Arrival pattern at stall.

Number of arrivals during a 30-second period	Frequency: number of 30-second periods in which the given number of arrivals was observed
0	49
1	26
2	18
3	10
4	10
5	8
6	7
7	6
8	5
9	4
10	3
11	2

During the last week 624 customers were served and 823 hot-dogs were sold at a price of 30p each. As usual, surplus cooked sausages were sold to the staff at 5p each and total takings for the week (including sales of surplus rolls) came to £251.33.

Although the stall is open only half-an-hour each day, staff employed have to work 1½ hours each day so that preparation and cleaning may be done. The hourly rate of pay for staff is £1.50, and staff are paid when on sick leave.

The data of Table 1 relate to the arrival pattern at the stall and were collected by students doing a statistical project. In the course of data collection it was also noted that arrivals tended not to join the queue if there were seven or more customers already there.

Find the best way of organising the stall – particularly the best number of servers to be employed.

26 PRODUCT PRICING (D: M. Wilson)

C. Threwes Ltd is a small expanding firm that manufactures delicate transparent objects. The firm is organised as four main sections which are the office, the glass shop, the plastics shop, and 'stores'. Details of what is done by each section have been obtained from discussion with the heads of each of the sections and are summarised as follows.

The glass shop employs twenty people at an average wage of £100 a week. Having made an item, a craftsman then records the time used for that job, on a job ticket. Material costs are negligible in comparison with wages. Each employee works a forty-hour week, but only about twenty-five hours of this appears on job tickets since staff tend not to record times used for design and material selection which takes the form of discussion with workmates and customers.

The plastics shop holds ten expensive machines, and thirty people are employed there. The average wage is also £100 a week. As with the glass shop, craftsmen log time spent on jobs on the job tickets. This recorded time accounts for about 63% of their 40-hour week. Again material costs can be neglected.

The 'stores' are somewhat misnamed in that they cover a number of functions. These include security, cleaning, machine maintenance, and the obtaining and storing of raw materials. Their weekly wage bill for the twenty-two staff (which includes part-time workers) is about £750. The section estimates that about 200 man-hours a week are spent on problems directly associated with production, this work being evenly split between the glass and plastics shop. In addition, about 50 man-hours' work is needed for tasks directly associated with the office, including regular polishing of the Sales Director's car!

The office also covers a number of functions. One of these, namely accounts, is poorly performed. It is unwieldy owing to the company growth and is run by a man who is rather inept in this field. The office estimate that they spend 210 man-hours a week on work directly associated with the production departments

– about two thirds of this for the plastics shop if one operates on order volumes. Work for 'stores' is also estimated at about 140 man-hours each week. (These estimates are vague as they had never been required before.) The weekly wage bill (plus director's fees) is £1050 in round terms.

As far as can be determined fixed costs are about £320 a week for each section (office apart) and in addition, stores use about £80's worth of fuel, oil, and petrol each week. The costs for the office, which include many miscellaneous items, amount to £560 per week. There are also die rentals and depreciation to consider for the plastics' machines, and these work out at £1100 a week.

At the moment the price of an item is worked out by taking the hours recorded on its job ticket and charging this at £8.50 per hour. There is a pay rise due that will affect the office and about 50% of the 'stores' staff. This will increase the wage bill in those sections by 12% and 5% respectively.

Taking, as an example, a job ticket with four hours glass shop work recorded on it, then what did it cost the firm overall to make, and what effect will the pay rise have on this cost? What would you suggest as a fair price for the item?

Index